The Natural Laws of Plot

ALEMBICS: PENN STUDIES IN LITERATURE AND SCIENCE

Mary Thomas Crane and Henry S. Turner, Series Editors

A complete list of books in the series is available from the publisher.

The Natural Laws of Plot

How Things Happen in Realist Novels

Yoon Sun Lee

PENN

UNIVERSITY OF PENNSYLVANIA PRESS

PHILADELPHIA

Published by
University of Pennsylvania Press
Philadelphia, Pennsylvania 19104-4112
www.upenn.edu/pennpress

Printed in the United States of America on acid-free paper
10 9 8 7 6 5 4 3 2 1

A Cataloging-in-Publication record is available from the Library of Congress

Hardcover ISBN: 978-1-5128-2340-0
eBook ISBN: 978-1-5128-2341-7

For Ji Min and Ji Hoon

CONTENTS

CHAPTER 1

Novels, Novel-Theory, and the History of Objectivity

In the context of novels, the word *objectivity* will evoke for many a way of telling the story with a certain detachment or from a third-person point of view. But the objectivity that this book will examine does not characterize a mode of narration. Rather, I want to look at objectivity as something that has to be built into the plot of a novel. Here is a curious description of what that might look like, from an essay by the late eighteenth-century novelist and philosopher, William Godwin. Godwin is comparing the relative advantages and disadvantages of romances (a category in which he explicitly includes novels) and histories. Novelists or romancers can understand their characters more profoundly because they have created them. They can show how "such a character acts under successive circumstances ... how character increases and assimilates new substances to its own, and how it decays, together with the catastrophe into which by its own gravity it naturally declines."[1] Notice how Godwin talks even about invented character as if it were an organic substance or solid projectile whose motions can be weighed and calculated. But when it comes to plot, historians, he argues, have the advantage because the objective connections between events are too intricate, too many, and too small, for any novelist to be able to portray them in a plot in any realistic way. "Naturalists tell us that a single grain of sand more or less on the surface of the earth, would have altered its motion, and ... diversified its events." The historian is more fortunate because "the events are taken out of his hands and determined by the system of the universe.... The romance writer, on the other hand, is continually straining at a foresight to which his faculties are incompetent.... That principle only which holds the planets in their course, is competent to produce that majestic series of events which characterises flux, and successive multitudes."[2]

In this striking final sentence, with its echoes of Newton, Godwin draws our attention to plot as the representation of how things happen and cause other things to happen. It is a chain of events that could include items as small as a grain of sand. Yet that "series of events" is the opposite of random or arbitrary. From the grain of sand to the fate of nations, the way that one event is linked to another depends on "the system of the universe . . . that principle . . . which holds the planets in their course"—on the working of gravity or of something even more objective, penetrating, and far-reaching.[3] Though Godwin's views on necessity were rather extreme, what he says about plot here should not be dismissed.[4] As the title of one of his novels suggests, he believes that novels should represent *Things as they are*.[5] In a realistic plot, things have to happen and cause other things to occur in ways that appear consistent and uniform rather than arbitrary. But this is not a simple matter.

This book will consider plot as an unfolding series of narrated or implied events and situations, together with the conditions that make them possible. I will focus on plots in novels rather than in drama or other genres because, as I hope to show in this chapter, novels have to rely more deeply and intricately on an environing world to carry along any action, to bring it from a wish, hope, or aspiration to something objectively realized. Of course, the difference between a mere intention and one that is accomplished is not unique to novels; nor are novels alone in attending to the difference between false beliefs and true ones, or delusive appearances and actuality. In that sense, objectivity forms part of any fictional world or plot. It is included within the very concept of an event.[6] Plots assert that some things happened or existed, while others did not.[7] We can think of objectivity, then, as a kind of fold or doubling-down of fictionality. It is part of what makes a fictional narrative closer in some ways to a false claim than to a dream that makes no pretense to describe what actually happened.[8]

But if lies need some kind of connection to the truth in order to be plausible, however exiguous that connection may be, the plots of realist novels also have to rely on models of the external world in order to stand up to scrutiny. One of the claims that this chapter will make is that plots are more than free-standing, self-enclosed structures that can be analyzed as form, as logic, or as literary inheritance. Those categories do not exhaust all the aspects of plot in the novel. In the language of narratology, this book will focus on story, the events that occur. But I will not sift the story's events out from the discourse, understood as the consecutive verbal embodiment offered by the text. My approach will depart from narratological practice in some other ways as well.

Recent narratologists have focused productively on the concept of storyworld, defined as a "mental model of who did what to whom, when, where, why, and in what fashion in the world to which recipients [of narrative] relocate."[9] Philosophically, the storyworld has been defined as an "alternative possible world."[10] It has been described by cognitive narratologists in terms of codes embedded within narratives that readers decipher and use to build up the sense of a particular world in their heads as they read. Like these narratologists, I see plot as inextricable from a world in which its events occur. But I will suggest that the storyworld is not something constructed exclusively by an individual narrative text or reader. Nor does it depend primarily on universal structures of human cognition or embodiment. Plots depend crucially on the act or stance of referring to an external world, one that can be investigated and also imagined to exist apart from both story and reader. The qualities and contours of that external world are as complex as its relation to the fictional one.

This book explores how plot works in the novel as this genre comes to be read and named as such over the course of the eighteenth century in Britain. I will focus on novels that claimed to represent the world realistically or objectively; by 1815, to a large extent that claim became an accepted part of the genre's self-definition.[11] I will look at how ideas about knowledge, experience, causation, and event develop and change during this period. In the following chapters, I argue that plots in these novels build on various models of how the world works, of how things in it move and behave. Events occur in a world of intermediate causes: events have particular material causes, apart from whatever meaning or significance they may possess, and these causes are taken with great seriousness. A well-known example is the footprint that Robinson Crusoe finds on the beach one day on his apparently deserted island. Even if providence can be surmised to operate somewhere far back in the chain of causes, there still had to be a material foot that pressed into the sand, at a particular moment between tides. We might say that in realist plots, things have shadows and textures, as well as depths and surfaces; events have hidden faces, and substances, as well as movements, exist only for limited durations. Plots require real footprints, and footprints can only be left in certain conditions, as any reader of detective fiction knows. Events in these fictions depend logically, if somewhat paradoxically, on a real world of physical forces and laws: a world that exists apart from the fiction.[12] In other words, and to make an obvious point, we look for realism not only in the way a novel is narrated, but in the way that events happen or do not happen.

Realist plots incorporate other ways of knowing and investigating the world. Their storyworlds require a kind of coherence vouched for from outside the novel. Novels have long been understood as mixed, heterogeneous genres; I want to suggest that their plots or stories in this period were also mixed up with models of the physical world that stem from elsewhere.

That elsewhere, as this book will suggest, begins in the broad and sprawling field of natural philosophy, an investigation of the natural world that likewise sought to answer questions about how and why things happen, how things can be made to happen, and what kind of value an event may have. I will mostly use the term *natural philosophy* rather than *science*, in keeping with the usage of the period; its projects ranged from Newtonian physics and optics, botanical classification, and chemical experiments to medicine and physiology. It was carried out in academic, public, and private settings, by men and women of varied social classes; and it was disseminated in a vast range of publications, from the *Philosophical Transactions of the Royal Society* to monthly magazines, to narratives for and about children. I will suggest in the following chapters that natural philosophy, far from being a rival, was closely enmeshed with the way that novels conceived and executed their plots.

Earlier accounts of the novel's relation to natural philosophy emphasized the importance of firsthand observation or experience in the type of enterprise pursued by the Royal Society, along the lines recommended by Francis Bacon.[13] Detailed reports of firsthand experience acquired a new kind of legitimacy and provided an opening for a new genre to emerge. In keeping with my focus on plot rather than narration, though, I will argue that the novel and natural philosophy shared even more than a way of reporting or a preference for firsthand experience. Both demonstrated a keen interest in the concept and the value of objective experience: in questions of how things happen, how far causes can be traced, and how significant the circumstances of an event may be. What is the value of an event that happens only one time? Is there an important difference between an event that happens by itself and one that was deliberately arranged? Even if an event's circumstances were contrived artificially, can it reveal something that is true everywhere and always? Such questions cannot be answered solely through the discourse of probability. They require attention to all the circumstances of an event, to all the abstractions that can be brought to bear on the bodies involved in it, and to the kinds of knowledge that can be drawn from it, or not. We can see in the development of what Peter Dear calls the event-experiment how close the concerns of natural philosophers were to those who thought about fiction, and about plots in particular. "An 'experience' in the Aristotelian sense was a statement of *how things happen* in nature, rather than a statement of *how something had happened* on a particular occasion: the physical world was a concatenation of established but sometimes wayward rules. . . . But the experimental performance, the kind of experience upheld as the norm in modern scientific practice, is . . . sanctioned by reports of historically specific events."[14] In this same period of the seventeenth and eighteenth centuries, Dear argues, the event-experiment establishes itself as a decisive form of knowledge and displaces the more general idea of experience as that which happens all the time, or most of the time.

Like event-experiments, novels place in a new light the familiar Aristotelian distinction between history, which tells what happened one time, and poetry, which tells what could or would happen every time. Aristotle asserts that "the universal that poetry aims for" is "the sort of thing that (in the circumstances) a certain kind of person will say or do either probably or necessarily."[15] In between the universal (something that happens in the same way every single time because it has to happen that way) and the singular or particular (something that happened one time), we find the entire world of novels' plots, their iterations and reiterations, the fundamental, objective uncertainty on which they rest, and their search for a kind of knowledge. The plots of novels suggest that few events fall unproblematically into the category of the universal or of the singular.[16] They often hinge on those circumstances that Aristotle seems to dismiss so lightly. An action is repeated: does this mean that the third time it'll happen the same way? What if the circumstances are different? Does it matter if the circumstances were spontaneous or artificially arranged, brought about through someone's agency? Does it matter if someone was there to see or hear it? What counts as experience in a plot? What counts as experiment? When is an experience or event conclusive? How do you know whether the results depended on the local circumstances or on the qualities and characteristics of the thing itself?

The plots of novels describe particular events, bodies, motions, and behaviors. At the same time, as I will try to show, they invite you to imagine these at an abstract level: as matter in motion, as natural types or species, or as substances that can unite and dissolve, but not at will and not randomly. The novels I will examine in the following chapters appear to rely eclectically on various accounts of these phenomena in order to invoke the sense of a unified, uniform, yet not fully transparent universe, one that behaves according to its own laws. They also rely on the modes in which knowledge was produced or recognized as such: procedures, protocols, notions such as the event-experiment. I am not arguing that novels' plots themselves produced or counted as a type of natural knowledge or a direct way of knowing the world. But they can perhaps approach the status of a postulate. A postulate was defined by seventeenth-century mathematicians "as a supposition that something can be done or brought about . . . assuming or affirming some evidently possible Mode, Action, or Motion of a Thing."[17] In order to affirm that some action or motion of a thing was possible, novels could not simply describe something that existed or happened one time.[18] Nor could they rely exclusively on statistical thinking to decide what was "evidently possible."[19] The latter required evidence of some other kind.

Plots had to incorporate or acknowledge somehow the conditions of an event: all the externalities that determine whether an action realizes itself or fails, whether an event reveals something or does not. And those externalities required a certain coherence at an abstract level; they required a form of thinking that could assimilate and account for the meaning of particular bodies or events. This is where I see natural philosophy and the novel converging in the eighteenth century. Both continued to explore questions of what experience could be said to reveal. If older models of natural philosophy "explained why things happen in terms of the essential natures of bodies," while newer ones gave more importance to "how something had happened on a particular occasion," novels' plots benefit by refusing to commit themselves fully to either.[20] They also benefit from an ambiguity that verisimilar plots possess: do such plots simply offer "observational perception of nature's ordinary course" (with perhaps an accident or two included), or do they "by design [subvert] nature" in order to produce some postulate about the world?[21]

This book asks what the plots of realist novels do. It tries to read them in new ways, and it will look at an aspect of plot not often examined: what we might simply call the background. Among contemporary students of narrative, plot is often thought of in terms of what Jerome Bruner called canonicity and breach: a plot exists when "an implicit canonical script has been breached . . . in a manner to do violence to . . . the 'legitimacy' of the canonical script."[22] In the simplest terms, we believe that certain things happen in a certain way; when that belief is disrupted by some event, only then do we have a narrative, or a plot. But to focus only on the breach, on the unusual event, is to miss in a way the larger, deeper, and more powerful dimension of plot: the way in which it does in fact refer to the world and requires a fairly developed sense of how the latter works. The components of narrative, including the names of places and events that happen, can become what Bruner calls "an invented referent," mixing together the fiction's internal preoccupations with aspects of the real, external world.[23] But Bruner also provides the remarkable phrase "ordinarily mundane event" to describe how events in fictional plots rely on an implicit sense of everydayness, and how they can shake up as well as affirm our intuitive sense of how the world is composed.[24] This book will focus on that sense, or the "script" without which a breach cannot occur or be distinguished as such. As a result, what might be dismissed by some critics as "filler" becomes much more important in my argument. And indeed, the distinction between description and plot will sometimes be handled not as the difference between two types of positively identifiable textual components, but as a difference of event-scale, tempo, timing, and explicitness of causal relation.

While I will offer a range of answers in the following chapters, I assume throughout that the plots of realist novels take up in some fashion the question of order or regularity in the external world as well as the question of how that regularity can be known or represented. To explain why, I turn now to my reasons for using the term *objectivity*. Through this concept, I want, first, to draw attention to a certain quality that realist plots attribute to events that happen and to the way in which they happen. Plots obviously include human characters and social and historical as well as natural events, and events include utterances and discoveries, as well as movements individual or collective. But in a novel that can be considered realist or naturalistic, all these events are integrated into a natural world and require an infrastructure of material circumstance in order to occur. For example: someone or something falls, or misses its intended target, or simply changes over time. If we ask why, the answers do not lead back to anyone's intentions, plans, purposes, or desires. Fate, destiny, or providence may offer a semblance of an answer, but not a sufficient one for the realist novel. Rather, plots make reference to the way the world is set up, to permanent and apparently universal baseline conditions within the novel's world that are required for something to happen, or that keep something from happening. While plot accommodates accidents and elements of chance, the latter depend on a sense of the fundamental regularity of the natural world.

Another way to put this is that in the plots of these novels, there is a quality that causation possesses that allows it to be distinguished from volition or even human agency. It appears to involve an entire system of things, such as Godwin describes, or, at the very least, a chain of related events that depend as much on the properties that things possess as on the intentions of actors.[25]

Even events that seem to exist purely on the level of the social happen only at the end of a chain of small intermediate causes that, we are meant to understand, sometimes act on their own. Emma Woodhouse, Austen's character, expresses this insight when she gives credit to rain-showers, umbrellas, bootlaces, and other such matters (as well as to herself), for bringing about marriages.[26] Austen's plots take pleasure in emphasizing the speed or slowness of carriages, the thickness of a necklace chain, the ripening of strawberries, the heat of the sun, or the width of a room in which bodies will make contact or not. Such phenomena play a role in determining whether a social event will occur, and what its consequences will be.[27] The idea of causation can subsume the distinction between human and nonhuman, natural and social, physical and psychological. The same can be said of the category of force. Plots can weaken while yet upholding the distinction between internal and external states, and between contingent relations and necessary laws.[28]

Sometimes, at the end of a novel, the plot seems curiously compelled to account for itself, as if the plot could not be complete without turning around and exhibiting itself as a continuous, objective causal chain. At the end of *Pride and Prejudice*, for example, there occurs between its two main characters a long and detailed discussion of cause and effect, the process and pace of internal change, and the external means that brought about their union. Together they trace the causes of various phenomena they have observed in others (delight, disappointment, gravity, thoughtfulness) or the effects of Lady Catherine's interference ("exactly contrariwise" to her intentions); and the novel's final words refer to the characters who had played the part of intermediate causes, "the means of uniting them."[29] Yet novels have to be careful about attributing objective causal influence too boldly. The enterprise is risky, as when Radcliffe lifts the veil at the end of her *Mysteries of Udolpho* to reveal the material causes behind various phenomena: secret passages in the walls, smugglers, a deranged nun wandering through the woods with a lute. These causes produce disappointment because they retain social forms; they do not seem grounded deeply enough in the objective characteristics of the physical world. As a later chapter will argue, the novels of Radcliffe and Burney locate more satisfying explanations for why things happen in the intimate substances, textures, and composition of the physical world, which includes its human characters and even their perceptions.

In its investigations, eighteenth-century natural philosophy handled causal explanation carefully, leaving some areas dark, while offering models of certain kinds of physical behaviors.[30] Forces that were too large, too small, or too ethereal could only be surmised, though their effects could be described, often in terms of the movement of bodies. In the novels of this period, too, causality can be deeply embedded; it can be shifted onto the physical or material composition of that world, as the example of Defoe will show.[31] At the same time, we should note that this composition often emerges only through the way it is described in the discourse or text—as in Radcliffe's extraordinary assemblages of light, sound, and time. We cannot get at it by abstracting and labeling events according to a predetermined ontology of plot that relies largely on social categories (such as courtship or marriage).

So far, I have used *objectivity* to refer to a certain quality that novels' plots try to lend to the way that things happen, and to a world whose structures and natural laws can be revealed by events or experiments. But the concept of objectivity is neither static nor timeless. It possesses its own history, which is essential to the argument of this book. I will suggest that the history of objectivity offers a frame within which we can situate the developments of natural philosophy, the novel as a genre, and novel-theory, or attempts to describe or define this genre. These developments appear to share the same fulcrum: the late eighteenth and early nineteenth centuries. Though later chapters will pursue this history in more detail, I want to summarize two accounts of a shift that occurred in how the knowledge of nature was understood and produced. I will then trace their consequences for how we understand the novel's history, as well as theories of the novel.

In their history of the concept, Lorraine Daston and Peter Galison argue that objectivity was not defined as the antithesis of subjectivity until close to the end of the eighteenth century. It was understood that creators of the scientific images used in anatomy, botany, or zoology atlases needed to rely on their own accumulated experience, their knowledge, skill, and instincts in order to produce true images of objects. They had to erase the defects or ignore the idiosyncrasies presented by any particular specimen, and show it either in its "ideal" form or in a "characteristic" one, which "locates the typical in an individual."[32] Daston and Galison argue that this attitude underwent a significant change in the nineteenth century. Objectivity came to be seen as something attainable only if the self were actively suppressed or even replaced by a mechanical means of producing an image. By the middle of the nineteenth century, subjectivity was identified as "the enemy within," a threat that had to be overcome: "Objectivity meant cultivating one's will to bind and discipline the self by inhibiting desire, blocking temptation, and defending a determined effort to see without the distortions induced by authority, aesthetic pleasure, or self-love. . . . the regulation of interior states and external procedures defined objective vision."[33]

Mechanical objectivity is the name that Daston and Galison give to this newer epistemic virtue: "knowledge unmarked by prejudice or skill, fantasy or judgment, wishing or striving."[34] We can already note the potential connections that could be made with novels, and especially the plots of novels such as Charlotte Lennox's *The Female Quixote* (1752), Maria Edgeworth's *Harrington* (1817), or Jane Austen's better-known *Pride and Prejudice* (1813), all of which will be discussed in later chapters. The threats to objectivity mentioned are the very forces that propel these plots: prejudice, fantasy, striving, authority. While critics have given plentiful attention to these forces, they have often considered them in terms of ethics, politics, or individual moral development. But they are also implicated in the history of natural knowledge, of how the external physical world is organized and composed, how it moves and behaves.[35]

Kant's model of the subject, "dynamic and autonomous . . . projecting itself outward," contributed to this change, Daston and Galison argue. Unlike the "passive and permeable self" of Enlightenment psychology, the self that was described by Kant's critical philosophy threatened to be too powerful, "overactive and prone to impose its preconceptions and pet hypotheses on data."[36] Rather than trying to pin down the causes behind this shift, I want to invoke another account of it, one whose relevance to the history and theory of the novel I will stress throughout this book. Michel Foucault's *The Order of Things* concurs in locating a threshold at the end of the eighteenth century. While Foucault does not use the term *objectivity*, his description of knowledge in the late seventeenth and eighteenth centuries emphasizes a similar (to us, strange) lack of concern with the placement or function of the knowing subject. Natural histories exemplified how knowledge was conceived and practiced in what he calls the classical era: beings could be gathered, arranged according to their identities and differences, seen, and therefore known. Representation, therefore, or the practice of making beings visible, was not regarded in the classical era as suspiciously tainted by subjective will or bias, or other limitations inherent to subjectivity. Representations existed in the same way that beings themselves existed.

In the latter part of the eighteenth century, according to Foucault, a change happened. If the classical era had relied on a concept of order as the form of knowledge itself, the modern era that begins around this time found its attention engaged by a very different concept of history, "that radical mode of being that prescribes their destiny to all empirical beings."[37] To know an organism was no longer to place it in a clearly visible array of similar or slightly different fellow-beings, but rather to look for its internal structures and how these functioned in the history of the species as well as the individual. The act of representation became newly problematic because the human subject who observes and represents, who seeks to know and understand, emerged as a central problem: "The space of order, which served as a common place for representation and for things . . . is from now on shattered: there will be things, with their own organic structures, their hidden veins, the space that articulates them, the time that produces them; and then representation, a purely temporal succession, in which those things address themselves (always partially) to a subjectivity, a consciousness, a singular effort of cognition, to the 'psychological' individual who, from the depth of his own history . . . is trying to know."[38] As this account suggests, Foucault also sees Kant's critical philosophy as playing a crucial role in this shift; it closed down one metaphysics, he argues, and opened up others more intimately tied to subjectivity: "those philosophies of Life, of the Will, and of the Word, that the nineteenth century is to deploy."[39]

To recapitulate the story told by these historians: toward the end of the eighteenth century, the human subject who assembles knowledge and represents the world becomes the object of a new, anxious scrutiny.[40] These are also the decades in which science undergoes differentiation into disciplines, literature and science start to be seen as divergent enterprises, and the novel begins to become a serious cultural institution.[41] As the knowing subject became a problem for knowledge, the novel emerged in tandem with a discourse about this very genre, its origins, its problematic effects, its affordances and limitations. In other words, arguments about the novel's rise were part of the phenomenon they claimed to describe. The novel was a problem for itself from the beginning. In her 1785 *Progress of Romance*, Clara Reeve describes how, by 1752, the year that Lennox's *The Female Quixote* was published, "the press groaned under the weight of Novels, which sprung up like Mushrooms every year . . . ten years more multiplied them tenfold . . . till they became a public evil."[42] Reeve shows a late eighteenth-century world in which, simultaneously, readers clamored for novels, "Manufacturers of Novels were constantly at work," and "Reviewers complain[ed] bitterly of the fatigue of reading them."[43] And all involved tried to put their fingers on what made the novel different, new, and problematic.

Novels participated in the cultural shift that Daston and Galison describe; so too did the theory that accompanied this genre like a shadow. That criticism suggested that novels could and should possess a certain distinctive objectivity. Clara Reeve defines the genre thus: "The Novel is a picture of real life and manners, and of the times in which it is written. The Romance . . . describes what never happened nor is likely to happen.—The Novel gives a familiar relation of such things, as pass every day before our eyes, such as may happen to our friend, or to ourselves; and the perfection of it, is to represent every scene, in so easy and natural a manner, and to make them appear so probable, as to deceive us into a persuasion (at least while we are reading) that all is real."[44] The novel is a picture of what happens all the time, "every day before our eyes." Notice that, in Jerome Bruner's terms, the novel's plot offers the canonical script; no mention is made of the breach. Reeve's formulation stresses its character as a representation, something made visible to the mind's eye as well as the physical eye. As a representation, it seems to conform to the concept of "truth to nature" that Daston and Galison attribute to the eighteenth century: it doesn't show what happened on one particular day, but instead offers a typical or idealized image of the everyday that's still accepted as objectively true.

The novel shows how things happen canonically: the emphasis falls on plot rather than on portraits of individual characters. Though the plots of novels don't claim to constitute natural knowledge in the same way as natural philosophy, as representations they cannot be kept entirely apart from this larger work. Representations bear knowledge in an era in which visible order counted for so much. They become suspicious in a moment when codes of knowledge begin to change. Novels' plots participate in that work of ordering, of establishing laws, relations, chains of connection, as my chapters will suggest. They move, sort, and sift bodies, measure and compare forces, even acceleration, as in Fielding's works. Yet novels' plots, especially as discussed by critics in the later eighteenth and early nineteenth centuries, also provide an opportunity to criticize the limits of representation, to point out the ways in which the latter may be "merely subjective" as opposed to "objectively valid."[45] Plots can come to be seen as instances of an overactive, meddling authorial subjectivity, one that assembles things according to its own wishes rather than the laws of the external world that they should reflect. We can already see a certain dissatisfaction emerge at the end of Reeve's definition: the plots of novels "deceive us into a persuasion (at least while we are reading) that all is real." Their objectivity can turn out to be a temporary illusion.

A particular kind of discontent with plot hints at a changing definition of objectivity. Such a sense is expressed even more clearly in the preface,

"On the Origin and Progress of Novel-Writing," that Anna Laetitia Barbauld appended to her fifty-volume collection of 1810, *The British Novelists*.[46] Together with Sir Walter Scott, who edited the *Ballantyne's Novelist's Library* (1821–24) in addition to writing his own novels, Barbauld played a key role in establishing the contours and contents of the genre. As Homer Obed Brown notes, "The eighteenth-century novel was invented at the beginning of the nineteenth century."[47] First, Barbauld affirms that novels offer "knowledge of the world," and then pulls back abruptly from that statement. Any such knowledge has to be false, she states, because

> Every such work is a whole, in which the fates and fortunes of the personages are brought to a conclusion.... Every incident ... is introduced for a certain purpose, and made to forward a certain plan. A sagacious reader is never disappointed in his forebodings.... But in real life our reasonable expectations are often disappointed; many incidents occur which are like "passages that lead to nothing." ... In short, the reader of a novel forms his expectations from what he supposes passes in the mind of the author ... but would often guess wrong if he were considering the real course of nature.[48]

Plot is the focus of Barbauld's dissatisfaction. *Tom Jones*, in particular, had been widely praised for the way its plot found a use for every casual encounter on the road, every dropped pocketbook, discarded muff, or lawyer in a hurry. Barbauld recasts this practice as a striking weakness that afflicts the entire genre. She complains that novels' plots are false not because the incidents described didn't happen, but because (a) they present a self-contained, aesthetic whole and (b) they reflect the intentions of the author instead of "the real tendencies of things." In Fielding's novel, the reader foresees a happy ending, she writes, "Not from the real tendencies of things, but from what he has discovered of the author's intentions."[49] Barbauld articulates a powerful dissatisfaction not only with the many dead ends that life offers, but with subjectivity itself, which now seems to stand opposed to objective reality. But we can also see in her account an older sense of objectivity as something that inheres in a representation of nature, and a novel's plot as an example of the latter. Only now that plot seems too neat, too artificial, and at the same time, suspiciously subjective, the product of somebody's fantasy instead of a reflection of the way things really are and the real totality that encompasses and unites them.

What we find, then, in the period when the novel was becoming a more serious cultural institution, is that plot came to appear problematic in a specific way. If novels wanted to come off as realistic, and there seemed to be no dissent on this point, they had to avoid the appearance of being too well-plotted. Scott makes this quite clear, as Homer Obed Brown reminds us. For his *Ballantyne Novelist's Library* (1821–24), Scott followed Barbauld's example in including Fielding and Richardson (already singled out by Reeve as preeminent novelists), as well as Sterne, Smollett, and Radcliffe. He did not include Defoe in this collection, but had already edited *The Novels of Daniel Defoe* more than ten years earlier (1809–10). He was the first to do so. In his essay on Defoe, Scott notes that the experience of reading these novels is different. It is not driven by plot, by the desire to know what happens, he asserts. Instead, we want "to read every sentence and word upon every leaf, instead of catching up as much of the story as may enable us to understand the conclusion."[50] Scott describes an almost hypnotic experience of being caught up in "an appearance of REALITY" (original emphasis) that Defoe has "given . . . to the incidents which he narrates."[51] It does not derive from his homely language, or even from the "character of the incidents" themselves. Rather, this sense of reality arises at least in part from the way the plot is handled, or not handled: "The incidents are huddled together like paving-stones discharged from a cart, and have as little connexion between the one and the other. The scenes merely follow, without at all depending on each other. They are not like those of the regular drama, connected together by a regular commencement, continuation, and conclusion."[52] Scott illustrates this with an example from *Robinson Crusoe*: "All the usual scaffolding and machinery employed in composing fictitious history are carefully discarded. The early incidents of the tale, which in ordinary works of invention are usually thrown out as pegs to hang the conclusion upon, are in this work only touched upon, and suffered to drop out of sight."[53] We never hear anything again about Crusoe's older brother, "who, in any common romance, would certainly have appeared before the conclusion." What ever happened to Xury? Or Crusoe's father? These characters and incidents "vanish from the scene, and appear not again." Scott emphasizes that in most fictions, "the author . . . does not willingly quit possession of the creatures of his imagination, till they have rendered him some services upon the scene; whereas in common life, it rarely happens that our early acquaintances exercise much influence upon the fortunes of our future life."[54] In other words, Scott praises Defoe for having built into his plot those very dead ends, those "passages that lead to nothing," that Barbauld found missing from novels like Fielding's.

The problem of excessive subjectivity in the novel, then, seems to be associated in this historical moment not just with a willful character, but with the plot: more specifically, with a plot whose shape reminds us of the existence of an author controlling every incident and character, wringing every possible "service" out of them. Defoe is unusual in that he can just let things go. But the metaphor Scott uses, of "paving-stones discharged from a cart," suggests that there is more at stake. The image of a heap of stones is meant to convey plotlessness, the lack of shape or design. Yet it also invokes weight, solidity, gravity, a kind of necessity in the way each incident unfolds. That sense is heightened when we recognize an allusion to the carts in Defoe's *Journal of the Plague Year* (1722), which Scott also discusses in this essay; a notable feature of that work is how the plague seems to wax and wane according to its own laws. Its narrator simply observes and documents, as in the memorable scene when he goes to watch bodies being dumped out of a cart into a great pit. Events may not have a clear connection with each other. Yet neither are they purely random: they do seem to follow some laws. Indeed, only when they lack artificial connections to each other can events possess this quality of happening on their own rather than adhering to an author's plan.

Instead of a general plotlessness, we see something a little more specific emerge as something the realistic novel should try to achieve: events that appear to follow their own laws, to illustrate some feature or quality of the external world in an objective way. Scott was not the only one to find in both Defoe and Austen the ability to create this effect.[55] In Scott's 1815 essay on Austen's *Emma*, which will be discussed in more detail in a later chapter, he summarizes the plots of *Sense and Sensibility* and *Pride and Prejudice* as well as *Emma*, indicating the importance that plot still possessed as a feature of the novel. But the scene he chooses to give as a long extract is not one that seems related to plot in the sense of a shapely arc. It is the dialogue that occurs in chapter 12 of the first volume between Emma's father and her sister on their respective favorite apothecaries. In this scene, which begins and ends with gruel, we see two characters, each completely locked in their own sense of the world, rolling down their accustomed mental paths, pushing against each other and failing to make the slightest impression. Mr. Woodhouse and Isabella both end up believing exactly what they had at the outset. No one changes their mind on anything, the merits of gruel included, and we can, along with Emma, predict exactly where and how the argument will go. It is a remarkable group portrait of the limits of subjectivity, of human subjects as things that move down their accustomed paths at a predictable rate.

Scott claims to have selected this scene because of its "precision," which "reminds us something of the merits of the Flemish school of painting," but it also stands out as an example of a scene that has no consequence in the development of the plot.[56] It simply represents a discussion that leads nowhere, that can only be cut off artificially when Emma dramatically changes the subject of conversation. Scott says the same thing about Austen that he does about Defoe: that its interest "pervades the whole work," every page; it is "a story which we peruse with pleasure, if not with deep interest," and indeed with a deeper interest "than one of those narratives where the attention is strongly riveted, during the first perusal, by the powerful excitement of curiosity."[57] There seem to be two kinds of possible plot-interest, then: one in which, through the author's artful planning, we focus on discovering the relations between events widely separated in the novel's discourse. In the other, we are absorbed in the event itself, the more inconsequential the better, in the way its unfolding suggests something about the world or "the real tendencies of things," to use Barbauld's expression. This latter kind of interest seems to be peculiar to the novel, as Scott and others envisioned it at the beginning of the nineteenth century.

The ability of novelistic plot to evoke a kind of objectivity has remained an important topic in critical discussions of the genre. But perhaps we have lost sight of it, as the novel has come to be identified with the principle of subjectivity, and as our models of objectivity have themselves changed. Ian Watt's still-influential *The Rise of the Novel* (1957) stressed that the genre's "primary criterion was truth to individual experience—individual experience which is always unique and therefore new."[58] Watt placed the novel in the realm of the social and economic, and linked it to the ideology of individualism.[59] The idea of the individual became the basis of the genre's "novel" identity, as well as its link to social, economic, and religious history. Watt thought of plot in terms of a formal plan or structure, and equated objectivity in the novel with a detached or external point of view.[60] This emphasis on individual subjectivity took two important and contrasting turns in novel criticism in the 1980s with the publication of Peter Brooks's *Reading for the Plot* (1984), and Nancy Armstrong's *Desire and Domestic Fiction* (1987). Brooks shifts the emphasis away from the empirical individual and toward desire as a psychological universal with a powerful structuring efficacy. Desire functions as an "armature of plot," as "a force that drives the protagonist forward."[61] Desire transcends individual subjectivity to become the interiority of narrative as such. Armstrong, taking a broader and more skeptical view, used Foucault's history of sexuality to argue that "the discourse of sexuality is implicated in shaping the novel . . . domestic fiction helped to produce a subject who understood herself in the psychological terms that had shaped fiction."[62]

The idea of a plot that could reflect "the real tendencies of things" was dismissed by twentieth-century criticism of the novel, as subjectivity and objectivity were driven further apart, or the distinction between them canceled by structuralism. Many important theorists of narrative, including Vladimir Propp, Claude Lévi-Strauss, Claude Bremond, A. J. Greimas, and Tzvetan Todorov, emerge from an era that Daston and Galison describe as devoted to "structural objectivity." This project, beginning in the early twentieth century, rejected visual images and image-making in favor of seeking out and identifying "enduring structural relationships that survived . . . cultural diversity, psychological evolution, the vagaries of history."[63] In the form of linked pairs of binary oppositions, the semiotic rectangle, or on the model of transformational grammar, these theorists wanted to lay bare the unvarying structures that lay beneath or behind all narratives. That often meant doing away with plot and searching for other ways of dividing narrative into small units.[64] Or plot events could be regarded as surfaces that betray symptoms of deeper social contradictions that again can be summed up concisely.[65]

Where plot was considered, it needed to be reduced to formulas, contracted into the smallest possible compass so that it could be consumed logically, rather than seen, felt, or experienced over time, or referred in any way to the external world. A remarkable example can be found in Vladimir Propp's analysis of one hundred Russian folktales. Propp reduced the plot of all of them to thirty-one functions that occur in an invariable sequence. As he described these functions one at a time, he offered "(1) a brief summary of its essence, (2) an abbreviated definition in one word, and (3) its conventional sign."[66] The signs that he used included alphabetic characters (including Greek), superscript numbers, mathematical symbols, and up and down arrows. Here are the final three functions: "XXIX. THE HERO IS GIVEN A NEW APPEARANCE. (Definition: *transfiguration*. Designation: T.) . . . XXX. THE VILLAIN IS PUNISHED. (Definition: *punishment*. Designation. U.) . . . XXXI. THE HERO IS MARRIED AND ASCENDS THE THRONE. (Definition: *wedding*. Designation. W.)."[67] Propp listed under each function some of the variants it might take; for function XXX, we might find, "The villain is shot, banished, tied to the tail of a horse; commits suicide, and so forth . . . we sometimes have a magnanimous pardon (U neg.)."[68] The plot event was boiled down to a sentence, a word, and then finally a single symbol. Propp then combined these symbols into a single, magnificently intimidating quasi-algebraic formula that represents the plot of one tale.[69] Significantly, Propp argued that the events or functions, as he called them, are related to each other not contingently and not by the external world but by their own inner logic: "The sequence of events has its own laws. . . . Theft cannot take place before the door is forced."[70]

The question of what constitutes a grammar of plot or a minimal plot was pursued by Todorov, Gerald Prince, and others.[71] The goal of such analysis was to cut through the clutter of examples, to subdue the inherent redundancy of plot, and to minimize its connection to the physical world of experience. In his 1966 essay, "An Introduction to the Structural Analysis of Narrative," Roland Barthes classified the smallest units of narrative, and declared these to be ruled by logical relations instead of empirical ones. The important "kernel" functions in a plot are defined logically (they open up more than one possible avenue of action). They can be related to each other by actual logic when they occur within a sequence (for instance, knocking at a door and the opening of the door), or by a "post hoc, ergo propter hoc" fallacy when they span a narrative. "Everything suggests, indeed, that the mainspring of narrative is precisely the confusion of consecution and consequence, what comes after being read in narrative as what is caused by."[72] Barthes thus neatly disposed of both causality and referentiality; he dismissed in another essay as "the referential illusion" the belief that the novel offers a "pure and simple 'representation' of the 'real.'"[73]

Theories of plot, and of the novel in particular, can be located within the history of objectivity as a concept or ideal. But the twentieth-century treatments described above tend to overlook the fact that plot is more than a formal structure that can be abstracted from the novel or a code that can be cracked. Plots in a novel aspire to represent worldly experience in all its duration and opacity; they question what makes that experience possible and gives it the shape and the consequences that it has. If we look at the elements of a novel that make it long, that distend the plot and give it heft, we find the representation of a world full of forces, substances, and movements that sometimes further but more often frustrate the aims of characters.[74] But that represented world, importantly, has to possess a certain coherence and predictability. It has to allow the tracing of causal chains behind its varied surface phenomena. These include natural phenomena as well as social ones.

Indeed, the prevailing model of the phenomenon, as something that occurs through a chain of material causes, is a natural one, as the next chapter will explain. If we assume that all plot events begin and end only on the level of the social, we will overlook much of what gives plots that sense of objectivity that readers like Scott found so compelling.[75]

I want to suggest that novels became powerfully appealing in the eighteenth and early nineteenth centuries at least in part because they aspired to represent an objective world as a totality, something complete and ordered in itself. Their plots offered a means to grapple with questions of objectivity as a dimension of the world, as an aspect of how things happen. A branch of novel theory in this same period makes this very argument. In his lectures on aesthetics (1818–29), Hegel describes epic as "the form of objectivity."[76] Objectivity here does not refer to a neutral attitude, a third-person mode of narration, or the removal of an intrusive self. It has to do with the agency and structure of the external world. It names the outsized, even dominant role that the outer world plays in shaping how an epic plot unfolds. This idea emerges in the context of Hegel's tripartite generic system of epic, lyric, and drama. The content of lyric is "the inner world, the mind that considers and feels, that instead of proceeding to action, remains alone with itself as inwardness . . . [in lyric] there is no substantial whole unfolded as external happenings."[77] Epic is also contrasted with drama. In both drama and epic, but not in lyric, such inward intuitions or thoughts take the form of action. In drama, however, "The action is not presented to our vision in the purely external form of something that has really happened . . . [rather,] we see it actually present, issuing from the private will . . . of the individual characters, who thus become the center as they are in the principle of lyric."[78] But in epic, the center is not to be found in the private will or thought of a character, nor in the determination of intent, nor in the movement from intention to action, all of which are so crucial to dramatic suspense. Hegel's phrase, "the purely external form of something that has really happened," gets at the special quality of epic in his view: epic "presents what is itself objective in its objectivity."[79] It represents the kind of action that "can be brought before us only within an outspread world and demands the portrayal of that world in its entirety."[80]

Plot lies at the heart of the epic genre in Hegel's account. Yet epic plot cannot be separated, even to the slightest degree, from the world in which it happens. That world is far from a mute setting or static array of objects: it is full of complex forces that work on their own, usually to frustrate or delay the end the character aims at. The scope and shape of the epic plot result from something like a deadlock, a balance of internal and external forces. Epic reveals "the form of objective and definite events and deeds in which there is an equipoise between (a) the agent's will and his individual aim, and (b) the external circumstances and the real hindrances that these impose on him."[81] Hegel stresses that epic deals in events, not actions: "In the case of *action* everything is referred back to the agent's inner character, his duty, disposition, purpose, etc.; whereas, in the case of *events*, the external side too acquires its unimpaired right, because it is objective reality which provides both the form of the whole and also a principal part of the content itself."[82] An epic's plot must "allow to the external circumstances, natural occurrences, and other accidental things the same right as that which the inner life claims exclusively for itself in an action as such."[83] Natural occurrences such as wind, tides, storms, or, in Homer's case, gods, play as important a role in epic plots as the protagonist's choices and decisions. They give shape and form to the events that happen.

Hegel's theory of epic is articulated around the same time as the novel's institutionalization, a process that included, as we have seen, critical reflection on the novel's plot and how it could create a sense of things or events happening on their own. The connection between epic and novel had already become a commonplace; Barbauld opens her essay by noting, "A good novel is an epic in prose, with . . . less (in modern novels nothing) of the supernatural machinery."[84] And when Hegel discusses the active role played in epic by "an outspread world," he does not, as we might expect, emphasize descriptions or catalogues of objects—shields, weapons, or ships. Rather, drawing on the ideas of Goethe and Schiller, he emphasizes the shape of the plot.[85] In epic, unlike drama, "The chief thing is not the devotion of activity to their own end but what meets them in their pursuit of it. The circumstances are just as effective as their activity, and often more effective." The *Odyssey*, an obvious example, "develops in full detail . . . what hindrances are put in [Odysseus's] way. . . . All these experiences . . . occur . . . without the hero's contributing anything to them."[86] Epic "especially turns aside in many ways from the execution of the main purpose."[87] Its plot likes to delay, even to move backward. This closely resembles the definition of the novel that we find in Goethe's *Wilhelm Meister's Apprenticeship* (written in the 1770s and 1780s, published in 1796): "The novel must move slowly . . . the main personage must . . . in some way or another, hold up the progression of the whole. . . . But drama must move quickly and the character of the main personage must press toward the end. . . . The hero of a novel must be passive, or at least not active to a high degree; from the hero of a play we demand effective action and deeds. Grandison, Clarissa, Pamela . . . even Tom Jones are, if not passive, yet 'retarding' personages. . . . In drama, the hero . . . either clears obstacles or pushes them aside, or he succumbs to them."[88] In the plots of novels, the outspread world and the protagonist's interiority might be said to strangely defer to each other. Goethe's remarks suggest that the novel's protagonist neither clears away external obstacles briskly, nor gives in to them at once, but remains to a certain extent entangled within them, even fascinated by them. The novel's plot is shaped not so much by the protagonist's aims or decisive action as by some elusive quality that the world itself objectively possesses as a whole and in every part of it without exception.

It is important to note that Hegel locates epic in a distant past, tying the form to a stage of social development that, he argues, provided the experience of a simple, single world. He points out that in the contemporary world, things and institutions can seem *too* free-standing, too independent and thus inimical to poetic creation. They tend to appear as "an already cut-and-dried and independent state of affairs," as "a prosaic arrangement of things" rather than part of a living totality.[89] Hegel notes, "Our modern machines and factories with their products, as well as our general way of satisfying the needs of our external life, would from this point of view be as unsuitable as our modern political organization is for the social background required by the primitive epic."[90] Modern divisions of labor or power are not conducive to epic roundedness. But what he overlooks, and what the following chapters of this book will explore, is that the external physical world in all its force and complexity remained the focus of wonder and investigation in both the novel and natural philosophy. In the eighteenth century, natural philosophers were actively formulating the laws and concepts that could allow the physical world to be thought of as a totality, as the next chapter will show in more detail. And the plots of novels were exploring in their own ways the kinds of order that might lie behind and hold together the behavior and appearances of the external world.

But the consciousness of historical change could not and cannot be dismissed. It will provide in a way the terminus for my argument about the realist novel. I hope in this way to extend somewhat the boundaries of this latter concept. A particular sense of history works its way deeply into the plots of Scott's novels and plays a crucial role in later theories of the novel. What we see is a displacement of totality away from the natural laws of the empirical world and toward a certain understanding of history. History, thought of as a set of latent forces, supplies an objective coherence that cannot be discovered elsewhere in experience. Here, I would like to focus on later theories of the novel. We can observe this move toward history, for example, in the criticism of Georg Lukács. His early work, *The Theory of the Novel*, develops Hegel's argument, noting that "the passivity of the epic hero that Goethe and Schiller insisted on" had nothing to do with their individual personality or their choice not to act. It arose from the fact that the epic's plot didn't need to be shaped by anything they themselves did: "The adventures that fill and embellish his life are the form taken by the objective and extensive totality of the world; he himself is only the luminous centre around which this unfolded totality revolves."[91] Lukács envisions plot in both epic and novel in terms of the fit between the action and the outspread, environing world. In epic, there is a perfect fit with no gap, no remainder. In novels, however, plots show the lack of fit between the world that action aims at (or shrinks from, in the case of inaction) and the world as it really is. In *Don Quixote*, for example, "All that opposes the soul must come from sources which are completely heterogeneous from it. Thus action and opposition have neither scope nor quality . . . in common. Their relationship to one another is not one of true struggle but only of a grotesque failure to meet."[92] In other words, the "circumstances and the real hindrances that these impose on him," to recall Hegel's words, exist in an external world whose existence the character is unable to conceive or to participate in consciously.

But Lukács later comes to argue that the novel can in fact achieve "true epic objectivity," and he bases this claim on the way that novels' plots can evoke the peculiar movements of history.[93] *The Theory of the Novel* had suggested that the novel could locate a kind of totality in the pure flow of time as something that brings about objective change in the external world and yet is also richly experienced at the subjective level.[94] That was the only option he saw because there was no external dimension in which, as he would put it later, "a person [could] express himself immediately and completely through a deed" in the world of the novel's plot.[95] But in his later study of the historical novel, Lukács argues that plot need not concern itself with individual action or expression. It attains objectivity by showing how events almost come about. "Epic, in all its forms, presents the growth of events . . . its maximum aim is to awaken this convergence of man and deed in the work as a whole, which it portrays, therefore, *at most as a tendency*."[96] The realist novel can be counted as a form of epic because Lukács now endows plot with a complicated, elusive ontology. It occurs on multiple scales. Actors can be social classes rather than single individuals. Moreover, an event in the plot does not have to take shape as one deed, a sequence of deeds, or a single "great historical collision," but can unfold instead through innumerable minor trends and tendencies, as incomplete or interrupted movements that may get picked up again later. Plot consists of "manifold connection," "diverse private destinies," and in such ways it can represent a "total world" that encompasses even the most minor private tensions.[97] Lukács comes back again to Goethe, reiterating that "dramatic action moves rapidly forwards. . . . But to portray the whole environment of an action, including nature and society . . . the action [of an epic or a novel] must be based on retrogressive motifs."[98] In other words, if a plot shows only a protagonist's movement toward a goal, or if we choose to focus only on that, we fail to see "the whole environment of an action." But we are forced to take notice of that environment when action cannot reach its end. To become epic—to become realist—the novel's plot must not focus on a single line of action or even a single arabesque of desire, to recall Peter Brooks's metaphor. Plot should aim to disclose "the total world within the entire circumference of which [an action] moves."[99] Plot becomes a circumference indicated by the objective forces and movements of history.

This idea that plot can assume a quality of objectivity by registering the deep, invisible currents of history becomes the basis of one of the most important theories of realism in the twentieth century, Erich Auerbach's *Mimesis*.[100] Auerbach's argument and method are worth examining here because they provide a bridge between the novel and realism as a mode of representation, a relation that I have not yet explicitly addressed. What I'd like to focus on in Auerbach's conceptualization of realism is the role of plot and its relation to history. Like Barbauld and Scott and for the same reasons, Auerbach dislikes the causal concatenations of plot, plot as artificial shape or plan.[101] He deplores the practice of establishing clear links between cause and effect as the unmistakable sign of willful artifice. Plot in this sense, as illustrated by the kind of causal explanation given for the scar on Odysseus's leg, becomes its own totality, rounded and self-enclosed. Auerbach calls plot in this sense "an order which [life] does not possess in itself."[102] He prefers authors such as Woolf and Joyce, who "have discarded presenting the story of their characters with any claim to exterior completeness, in chronological order, and with the emphasis on important exterior turning points of destiny . . . the great exterior turning points and blows of fate are granted less importance" (548). Such turning points and the clear shapes to which they give rise are merely reflections of the author's will.

Auerbach's commitment to removing subjectivity shapes even his own methodology.[103] He suggests more than once that he samples his texts randomly, perhaps in the service of a certain epistemic objectivity.[104] But objectivity is more than an epistemic virtue for the literary scholar; it comes to describe the relation of plot to history and becomes the basis of the unique aesthetic experience that realism offers. Auerbach suggests that while plot, in the sense of a "planned continuity of action" (552), had to be surmounted in order for realism to triumph, an alternative dimension of plot had to assume a greater importance. Suspense, something that is often identified with plot and is arguably achievable only by means of plot, becomes essential. But realist suspense can only come about in a certain way. Following Hegel, Auerbach argues that Homeric epic uses the "'retarding' procedure" in its plot, slowing down or straying away from the main action. However, he argues that there are other epic works "composed throughout with no 'retarding element' . . . but, on the contrary, *with suspense throughout*" (5, emphasis added). We might assume that narrative delay generates suspense, but Auerbach sees it differently. The suspense that he finds in the Abraham story from the Old Testament arises from an extreme narrative economy, especially where causal explanation is concerned. "The decisive points of the narrative alone are emphasized" (11). He sees it as pure plot laid out almost in notational rather than mimetic form. There are no detours, no digressions, no explanations. The mimetic time that separates God's command from Abraham's response is of course significant. But because the causes behind these remain undisclosed, the story is "permeated with the most unrelieved . . . overwhelming suspense" (11–12).

Yet even the omission of narrative detail or causal explanation is not sufficient to create realist suspense. Indeed, suspense that arises merely from withholding knowledge from the reader, or from delaying the narration or explanation of an event, simply betrays the existence of plot as artifice or lie, visible shape or plan. In Auerbach's theory of realism, suspense is elevated and displaced onto history itself, conceived as "a development of forces" (40) that lies behind the plot's events. No longer merely psychological or private, suspense seems to be the objective substance as well as the phenomenological trace of the history in which we participate. "Suspense and development [are] the distinguishing characteristics of earthly phenomena," he remarks (198). Instead of enjoying clear causal sequences, then, we should feel somewhat in the dark when we read realist plots. We should "receive the impression that the movements emerging from the depths of the people" are coming from somewhere we can't see and going somewhere that we can only guess at.[105] Realist plots offer sudden irruptions of change.[106] They show "everything unresolved, truncated, and uncertain, which confuses the clear progress of the action. . . . The historical event which we witness . . . runs much more variously, contradictorily, and confusedly" (19). He reiterates this point: "Abraham, Jacob, or even Moses produces a more concrete, direct, and historical impression than the figures of the Homeric world . . . because the confused, contradictory multiplicity of events, the psychological and factual cross-purposes, which true history reveals, have not disappeared in the representation but remain clearly perceptible" (20).

It is not an absence of plot that characterizes the realist mode, then, but a *superabundance* of plot that yet remains only partially or intermittently visible. We might recall Foucault's account of the modern episteme that emerges in the nineteenth century, with its search for hidden relationships and structures. The "interior 'mechanism'" that functioned inside the organic being, inside languages and economies, could not be directly represented or seen, yet it accounted for observed shapes and behaviors. Foucault's description of "this secret but sovereign mass," "the dark, concave, inner side" of the visible, captures the status of history in *Mimesis* as something that pervades every instant with suspense in the realist work: history as a totality that cannot be fully seen behind the plot phenomena that express it.[107]

Auerbach uses the language of randomness to describe the relation between plot and this sense of history.[108] Woolf "holds to minor, unimpressive, random events" (546); Joyce and Proust, too, focus on "random everyday events" (548). Here, a "random event" means the type of occurrence that no one would expect to be very important in the course of a life—for instance, the measuring of a knitted stocking, as opposed to the loss of a throne or "a change of fortune breaking in upon man from without and from above" (318).[109] It appears again in Auerbach's summary of "the foundations of modern realism": "The serious treatment of everyday reality, the rise of more extensive and socially inferior human groups to the position of subject matter for problematic-existential representation . . . the embedding of *random persons and events* in the general course of contemporary history, the fluid historical background" (491, emphasis added). He notes that "the term random [is] here employed to designate people from all classes, occupations, walks of life, people, that is, who owe their place in the account exclusively to the fact that the historical movement engulfs them as it were accidentally" (44).

In the context of fictive plot, however, the term may well take on a meaning antithetical to its usual one. The seeming spontaneity of the random event must possess the greatest artfulness. It cannot be redeemed just by being woven into the larger shape of the story, in the way that Fielding uses accidents and passing encounters. Rather, the realist plot full of loose ends or dead ends has to keep them loose and yet hold out the promise of glimpsing through such gaps and redundancies nothing less than the forces of history at work. Woolf, Joyce, and Proust show that "in any random fragment plucked from the course of a life at any time the totality of its fate is contained and can be portrayed" (547).[110] In realist fiction, Auerbach writes, "Something new and elemental appeared: nothing less than the wealth of reality . . . in every moment to which we surrender ourselves without prejudice" (552). When we stop trying to look ahead or look back or tie things together, there arises the sense of history as a total, all-encompassing and purposeful movement, a larger and richer circumference.

When Auerbach ends his history of realism with the novel, he turns not to a realist or naturalist novel but to Woolf's *To the Lighthouse*. Woolf's novel actually illustrates quite well Todorov's model of plot as based on linguistic transformation.[111] Its plot can be summarized into pairs of sentences that relate the first to the final part of the novel: James wishes to go to the lighthouse/James goes to the lighthouse. Lily wants to finish her painting/Lily finishes her painting. This is the type of relationship through which, Todorov argues, any reader can intuitively grasp plot as a sense of completeness. The ending of Woolf's novel could not stress completion any more dramatically: "It was done, it was finished," the novel concludes.[112] At the same time, if we look at its full extent, Woolf's plot notably demonstrates that delay, backsliding, or deferral that many critics of the novel found characteristic of the genre. Because of external circumstances and natural occurrences, to recall Hegel's terms, the trip to the lighthouse is postponed not only for one night or day but for many years. Auerbach, as we would expect, plays down this larger pattern. He remarks that "the meaning of the relationship between the planned trip . . . and the actual trip many years later remains unexpressed, enigmatic, only dimly to be conjectured" (551). Instead, choosing (or happening on) a passage from the beginning of the novel, he explores how the present moment and the single consciousness both dissolve, as Mrs. Ramsay measures her unfinished stocking against her child's leg. Auerbach insists that what is at stake here is not subjective experience but something objective: "we are here after all confronted with an endeavor to investigate an objective reality, that is, specifically, the 'real' Mrs. Ramsay" (536).

Though Auerbach only alludes to it in passing, the well-known "Time Passes" chapter in Woolf's novel seems to offer a test case for his argument about plot, suspense, history, and realism. As it can also stand as a test or limit case for my own argument, I will look at it in some detail.[113] By describing the passage of time in an uninhabited house, withdrawing its human characters from the narrative, the chapter represents exclusively that "background" on which Auerbach places so much value. There are no human actions that occur in the foreground. And that background is indeed replete with suspense. Despite the lack of embodied witnesses or participants, the chapter is full of events that occur on multiple scales: ceaseless tiny changes that occur inside and around the Ramsays' house after its occupants leave. "The saucepan had rusted and the mat decayed . . . the plaster fell in shovelfuls" (141). Yet far from appearing inevitable or occurring in a steady rhythm, the scope and timing of these microscopic events seem open. They appear to possess an immanent significance that keeps tempting an unspecified consciousness "to assemble . . . the scattered parts" (135) into a whole. The narrative asks, however, whether "That dream, of sharing, completing, of finding . . . an answer, was then but a reflection in a mirror, and the mirror itself was but the surface glassiness which forms in quiescence when the nobler powers sleep beneath?" (138).[114]

The "powers" beneath the surface do not seem to be those of history. In the chapter, the kinds of events that usually constitute the plot—deaths, births, and marriages of characters—are enclosed within brackets in the text: "[Prue Ramsay died that summer in some illness connected with childbirth]" (136). Similar brackets enclose what seem to be historical events: "[A shell exploded. Twenty or thirty young men were blown up in France, among them Andrew Ramsay]" (137). By setting both fictive and historical, intradiegetic and extradiegetic events within such brackets, Woolf's novel suggests that all plot can be seen as a variety of metalepsis, as if to narrate any event in the story were to breach a fourth wall and to refer directly to some outside reality.[115] Historical and fictive events are both, for the duration of the chapter, moved to some other level that is not continuous with the rest of the representation.

What are the powers, the forces at work in this chapter, then? Is it the case that there is no plot other than the events set off in brackets? It seems to me, rather, that the chapter shows us plot in its pure materiality, viewed as if under a microscope or in the chamber of an air-pump: an experimental device that will appear more than once in the following chapters. In this case, though, what's been evacuated are human-scale actors. Harder to remove is the persistence of consciousness, of looking, wanting, and seeking meaning or wholeness. The chapter does admit air, "stray airs" (132): "certain airs . . . crept round corners and ventured indoors. . . . they entered the drawing-room questioning and wondering, toying with the flap of hanging wall-paper, asking, would it hang much longer, when would it fall? Then smoothly brushing the walls, they passed on musingly as if asking the red and yellow roses on the wall-paper whether they would fade. . . . How long would they endure?" (130) Light and sounds are admitted, but reluctantly; the lighthouse beam threatens to become too active a presence, behaving like a physical cause: "as the long stroke leant upon the bed, the rock was rent asunder; another fold of the shawl loosened; there it hung, and swayed" (136).

The power seems to belong to plot itself, to objective eventfulness as such. The question of how long a substance or situation will endure lies at the heart of plot, its tests, enquiries, and machinations. The chapter tests, plays with the relation between events that happen the same way every time, and events that happen only once: "Once only a board sprang on the landing; once in the middle of the night with a roar, with a rupture, as after centuries of quiescence, a rock rends itself from the mountain and hurtles crashing into the valley, one fold of the shawl loosened and swung to and fro" (134). Yet it insists on the coherence of this world, as opposed to the bracketed events, and of the single force within it that moves everything from "the trifling airs" (141) to the waves to the creaky limbs of Mrs. McNab: "there was a force working" (143). The type of suspense that attends the microevents of this chapter's plot brings to mind the interest that Scott had felt in the realist plots of Defoe and Austen, the desire "to read every sentence and word upon every leaf, instead of catching up as much of the story as may enable us to understand the conclusion." Woolf's chapter also seems to illustrate what Godwin had said about the intricate nature of plot: "that a single grain of sand more or less on the surface of the earth, would have altered its motion, and . . . diversified its events." The empty house "was left like a shell on a sandhill to fill with dry salt grains" (141). It decays, disintegrates until a certain moment is reached: "Now had come that moment, that hesitation. . . . One feather, and the house, sinking, falling, would have turned and pitched downwards to the depths of darkness. . . . If the feather had fallen, if it had tipped the scale downwards, the whole house would have plunged to the depths to lie upon the sands of oblivion. But there was a force working" (142). This account suggests a drama occurring within a single, unified world. The force does not come from another level but emanates from inside this one.[116]

Can these kinds of events be considered part of the novel's plot or do they belong only to the setting?[117] How can action and environment be separated?

How can the action of the environment be measured and described? As we will see in later chapters, the kinds of events and, more importantly, the models of change that we see in Woolf's chapter will be important to the ways in which earlier novels build up their plots. Physical forces, even chemical processes lie at the circumference of events that occur, tests that are made. In the eighteenth century, as the next chapter argues, we see the investigation of what Jacques Rancière has called "an impersonal life, which does not know of the distinction between subjects and objects, human beings and inanimate things."[118] In experimental philosophy, the external world was understood, investigated, and measured as matter, and found to be uniform in behavior and consistency, subject to the same laws everywhere. In Peter Hanns Reill's words, "Matter's essence was streamlined and simplified: it was defined as homogeneous, extended, hard, impenetrable, movable, and inert."[119] It was argued that all natural phenomena followed the same rules, imparting an objective unity to the world. Intermediating causes, contact and connection between bodies moving and pushing against each other, explained how things happened. The experiment as a newly formulated protocol for scientific investigation also gave broader meaning to a single physical event, and connected the particular to the universal in a different way. Drawing on popular Newtonianism, this chapter suggests that the thing-filled, motion-filled, sequence-dependent plots of novels such as *Tristram Shandy* rely on this way of seeing the physical world as unified, connected, and materially homogeneous. Events and movements could involve humans and yet occur and continue without the concurrence of anyone's will; this is where the plots of realist novels could be said to conceptually begin.

In Defoe's fiction, the subject of the third chapter, the behavior of matter—its independent resistance or movement—provides human action with an indispensable alibi. It shapes its contours, timing, and ergonomic drama. Action is a grappling with matter, an attempt to move it or to stop it from moving. Defoe's plots create a sense of objectivity by stressing that human actions can never be completed or fully narrated because they are so entangled with the material world on different scales of size and visibility. They reshape and recast action as force, and embed the latter within the world described and explained by experimental philosophy. At the same time, the recursiveness of his plots suggests uncertainty about the status of a single event. Defoe's characters work very hard to move things or to keep them in their places, but neither can be completed permanently or demonstrated conclusively.

Natural history was the branch of natural philosophy committed to describing and ordering the diverse phenomena found in nature, whether living or nonliving. Both natural history and the novel in the eighteenth century aimed to know natural things by representing them in an ordered way. In the fourth chapter, I argue that the novel draws on both aesthetic ideas about the unity of plot, and a commitment to the uniformity of the natural, physical world. Together, these allow plot to become a kind of medium for displaying, comparing, and ranking diverse empirical types. In ways that resemble the methodology that Buffon proposed for natural history, plots in Fielding and Richardson seek to make natural types visible and therefore knowable through their behavior, movements, and relations to each other.

In the last quarter of the eighteenth century, as the fifth chapter suggests, the plots of some novels appear less interested in tracking the application of external force to a thing or person than in testing substances and observing how change occurs within and between them, at a different and more intimate scale. Chemistry, which developed rapidly in this period, studied materials and substances not as homogeneous aggregates but as possessed of unique properties that revealed themselves in qualitative changes, when brought into contact with other substances. Tobias Smollett's final novel stands as a transitional example: its many mixtures can be mechanically separated. But in Burney and Radcliffe, plot becomes a vehicle for investigating the intimate substance of character, conceived almost in chemical terms. The processes of composition and decomposition, fundamental to chemistry, become the stuff of plot in Radcliffe's novels: character and setting, light, heat, and sound change, blend, and dissolve into each other. Such impersonal events even appear to dominate social or human-scale actions.

The novel does not, however, follow a single track of development; nor does natural philosophy, which remains richly heterogeneous into the beginning of the nineteenth century. In the sixth chapter, I argue that Austen's plots still rely on a conception of bodies as aggregates of matter. The persistence of mechanical philosophy can be detected in how bodies, whether individual or collective, collide, resist, and attract each other. We see the application of force to matter, the phenomenon of movement, the unceasing operation of time, and background processes of change, all within the represented world and as part of the plot. But her plots also suggest that the boundaries of bodies can be fluid and changeable: her plots shuffle bodies in different combinations and suggest other scales of coalescence between them. The ontology of Austen's plots is complex. Initial pushes and inertial movement account for much, yet changes in composition over time are also scrupulously traced. Objectivity in her plots is based on the duration and extent of material bodies, regarded as minute, shifting aggregations. There is clearly more than one way to look at plot in an Austen novel, and other scales of eventfulness besides the social one.

As noted above, a historical shift appears to occur at the end of the eighteenth century, when the principles of subjectivity and objectivity begin to emerge as mutually exclusive. If plot had functioned as a plane of visibility, a means of objectively representing the world, this starts to change, as the seventh chapter shows. Subjectivity manifests itself through bias and distortion. Thus, the condition of quixotism becomes normalized in the novel. The self-experimentation and public lectures of Humphry Davy, science-writing for children, and the novels of Maria Edgeworth, William Godwin, and Austen all emphasize how subjects tend to view the world distortedly, how they tend to collapse inward upon themselves. Plot, as it unfolds over time, is now given the task of drawing out hidden inner truths and revealing biases. Almost like a form of mechanical objectivity, plot carries out impersonal procedures that elicit the truth about subjects. It also produces a therapeutic suspense, an avid attention to how events play out in the external world.

But the fallibility of perception becomes the basis for a deeper realism that the final chapter explores. Once the novel begins to incorporate this wariness toward subjectivity as a source of bias, it acquires a different, more dialectical means of invoking its access to objectivity as a dimension of the plot. The historical novel builds on the inability of the human senses to distinguish one's own movement from that of the surrounding world. Drawing on the physical laws of motion, the physiological study of vertigo, and the affordances of the novel, the plots of Scott's historical novels center the ambiguity and confusion inherent in the experience of physical movement. And the object of its representation will likewise shift from nature to a concept of history. Through such means there arises in his novels a strong conception of history as an objectively occurring, continuous shifting of the ground under one's feet. It begins to fill the hidden causal space behind natural phenomena. And it offers a new field as well as a new ontological ground for plot. A brief epilogue on Zadie Smith's novel, *White Teeth*, explores what a historical novel looks like at the end of the twentieth century, and how plot remains tied, though stubbornly and skeptically, to objectivity.

Plot has often been neglected by critics, and the question of realist plot is sometimes reduced to its ideological impact on beliefs about social life and individual identity. But plot is one of the novel's most important means of representation. And perhaps the moment has arrived to rethink the concept of ideology and its relation to forms of representation. Plot is powerful because it can be so openly heterogeneous, even promiscuous, in the models and assumptions it draws on to reveal an objective world. The following book does not defend objectivity as an epistemic virtue; rather it offers a history of how objectivity has been and continues to be a crucial dimension of realist plot and the ways in which we think about it.[120] It sees the genre of the novel and natural philosophy as examining shared questions. What is the authority or significance of a singular event? Are there laws that regulate how things move and stop, how bodies are composed, how types exist? How do things exist, change, and happen?[121] Novels do not justify a philosophical position or make a scientific claim; but they frame events, causes, and consequences as if they were objective. This aspiration can draw on protocols external to the genre: ways of investigating and imagining the physical world, or the historical one, as a totality. Defending his own "less artificial [i.e., less artful] practice of arranging a narrative," Scott insisted that in the real world, "in the general course of human life itself," plot unity does not exist: "The hero's later connexions are usually totally separated from those with whom he began the voyage, but whom the individual has outsailed, or who have drifted astray, or foundered on the passage." Ultimately, "the winds and tides, which are common to the element which they all navigate" will determine who ends up where.[122] Realist plots make us wonder about the objective dimensions and movements of that environing element.

CHAPTER 2

Matter, Motion, and the Physical World of the Novel

In the previous chapter, I argued that a concept of objectivity helps to distinguish the novel as a genre and realism as a mode. The current chapter now turns back to consider how objectivity, considered as a dimension of the realist novel's plot, emerges from the intertwined histories of the novel and of early modern science. If we start, as many do, with the assumption that the novel tells the story of a human individual's growth or development, it would make sense to look to philosophers such as Locke, Shaftesbury, Hume, or Smith for models of how individuals acquire knowledge about the world or come to behave in socially sanctioned ways. The novel's most important relationship in that case may well be with Locke, as Nancy Armstrong has recently argued: "Enlightenment philosophy left off and fiction took over . . . [to] perform what the Lockeans could only theorize." By means of ideas and words, novels could actually execute the work of producing "a self-governing individual."[1] But let's consider an event from a mid-eighteenth-century novel—one, moreover, that Victor Shklovsky called "the most typical novel in world literature."[2] The five-year-old protagonist of Sterne's *Tristram Shandy* is about to relieve himself out of a window when the window-sash falls down: "So slap came the sash down like lightening [*sic*] upon us."[3] One body moves and makes contact with another: in this moment, no words (or very few) are involved, no knowledge, little understanding. Of course, this unfortunate incident does play a part in Tristram's development as an individual. Despite the digressions for which the novel is famous, that development is alluded to in various forms throughout the novel: in the narrator's intention to write "the history of myself," or through the ludicrously elaborate, unrealized plans for his education drawn up by his father.[4] Tristram himself refers to the window-sash event as one that interrupts his youthful development. But considered more closely, this development, which is summarized in just a single sentence, is questionable. At best, he imagines it, like his narrative, as a normative, hypothetical trajectory: "I began to live and get forwards at such a rate, that if [this] event had not happened . . . I had put by my father, and left him drawing a sun-dial, for no better purpose than to be buried under ground" (338). That is, he would have outstripped his father's plans. But the movement of the window-sash is brutally actual, undeniable. Does it make sense to look for the meaning of this event as opposed to its causes and consequences, which seem to exist in a different, nonsignifying realm? Or could those be the same thing?

The falling of the window-sash is presented in the novel as an objective physical event—as matter in motion—for which a chain of causes has to be found. It doesn't seem to be sufficient to attribute it to special providence, or to dismiss it as a mere accident, though both of those responses are also acknowledged. We still need to know the regular physical causes through which it came about. We soon discover in the novel's next chapter the proximate cause for the window's behavior: the lead counterweights and pulleys had been removed. But why had they been removed? To be melted down in order to cast miniature artillery for Uncle Toby's model fortifications. Why did Uncle Toby build model fortifications? In order, originally, to help him explain to other people the circumstances of the wound that he had received during the Nine Years' War. Why was he wounded? Because a stone fell off a parapet during the siege of Namur. Why did the stone fall? Because of gravity: the injury to his groin was "more owing to the gravity of the stone itself, than to the projectile force of it" (69). (And obviously, gravity is why the window-sash fell, too.) In a sense, this causal chain is complete and well-forged. Every effect has an antecedent cause. But in another sense, it explains very little. Its painstaking nature raises questions about the adequacy of causal explanation and even the *parameters* of causal explanation within a fictional narrative. The more we know about why the window-sash fell down, the more arbitrary or confounding the event appears in one light, even while it also becomes perfectly intelligible, considered as an event that happens in the external, extended world: an event that cannot be undone. Or at least, its lasting effects require a great deal of work to repair. As such, this moment illustrates what I've been calling plot objectivity.

But can a simple movement like this, the falling of an object in the physical world, be considered part of a novel's plot? Shouldn't only intentional human movements count? Jonathan Culler gives as an example of something that is *not* part of the plot the following sentence from James Joyce's story, "Eveline": "The man out of the last house passed on his way home; she heard his footsteps clacking along the concrete pavement."[5] The structure of plot, Culler suggests, consists of important, morally and socially fraught action sequences that are performed by major human characters: in this case, the protagonist Eveline's "initial decision to leave and . . . final refusal to leave."[6] Even the movement of a minor human character fails to count as plot.

Culler's view is broadly shared. In his essay, "The Thread of the Novel," for example, Jacques Rancière contrasts "the truth of life" with "the lie of the plot" and links the latter to hierarchical social divisions that separate important people, "individuals who conceive of great projects and set out to achieve them," from unremarkable people who lead ordinary lives.[7] Rancière suggests that the mere "empirical succession" of life failed to count as a plot until the modern novel in the hands of Flaubert and Woolf figured out a way to link "microscopic and ever changing sensory events" with the important choices and decisions made by characters.[8] But I will argue that the realist novel explored such interconnections in detail. The realist novel does not limit plot to the actions, decisions, or intentions of significant, named human characters. Nor are such human actions simply opposed by chance or by accident. Rather, the objective, regular behavior of matter in space and time is fundamental to realist plots and the way that the latter work themselves out. In the realist novel, actions, choices, decisions, or discoveries cannot be separated from a network of external, material things and especially the ways in which they move, change their course, or stay still: how fast or slow, with what force or acceleration, and with what apparent relation to human causation. Lockean epistemology, with its focus on the sentient individual subject, can only go so far in helping us understand the way that novels' plots operate and the sorts of feelings to which they give rise. At the heart of novelistic plot lies the dialectic between subjective impressions and intentions, on the one hand, and the apparent objectivity of the external world, on the other. Indeed, when Locke's *Essay Concerning Human Understanding* makes a cameo appearance in Sterne's novel, it is wrenched in a materialist direction. Purporting to explain Locke's theory of "the cause of obscurity and confusion, in the mind of man," Sterne extends a well-known metaphor for the mind into an imagined wrestling match with an all-too-finicky material substance. Here are all the ways in which an "inch . . . of red seal-wax" can fail to receive a clear impression: "If Dolly fumbles too long . . . till the wax is over-harden'd, it will not receive the mark of her thimble from the usual impulse which was wont to imprint it"; if it's too soft, "It will not hold the impression, how hard soever Dolly thrusts against it," and even if the wax is the proper consistency, she may apply the thimble "in careless haste, as her Mistress rings the bell," leaving a confused and inaccurate impression (78). Sterne turns Lockean epistemology itself into something like a series of bungled experiments in determining the time-dependent changing density of a material medium.

Novels show a singular interest in the outstretched world, in the "external circumstances, natural occurrences" that become inextricable from the plot.[9] This interest signals more than a relation to the older genre of epic. It links the novel's gradual emergence as a genre to the development of early modern natural philosophy. From the late seventeenth to the late eighteenth century, the movement of matter in the physically perceptible world was a subject of broad investigation, most notably in Newtonian science.[10] While important studies have examined the development of the life sciences in relation to the novel, especially later in the eighteenth century, I will focus in this chapter on how experimental philosophy investigated the world in terms of the movement of matter.[11] We need to look at how it developed its interpretation of nature: what it claimed, as well as what it did *not* claim, about the homogeneity of the physical world, its essential composition, its movements, and what can be said about their causes. A certain concept of objectivity was deeply embedded in its interpretation. In Steven Shapin's words, "All seventeenth-century mechanical accounts set themselves in opposition to the tradition that ascribed to nature and its components the capacities of purpose, intention, or sentience."[12] But the movement of things in the physical world did follow certain laws. These laws, which were few in number, could be described, demonstrated, and explained; movements of nonliving things at all scales became to a large extent predictable and explicable. To the extent that living bodies were also made of matter, similar rules applied to them. But there remained a level of causality that was not broached by science, from which science retreated, and this left an important opening for the novel.

Equally consequential for the genre, as John Bender has shown, was the way in which natural philosophy devised a new form with physical, social, as well as epistemological dimensions: what Peter Dear has called the event experiment.[13] The execution and description of experiments according to certain protocols became an important feature of the dissemination of the mechanical philosophy.[14] They gave a certain form to what might be called consecutive experience. I will set the elaboration of fictional plot in realist novels in this context of natural philosophy's development. The realist novel's orientation toward plot objectivity emerges from the spaces that the mechanical philosophy envisioned, measured, but also left only lightly sketched in. These empty spaces were conceptual, institutional, imagined, and also physically real. In other words, natural philosophy in this period was aware of the gaps as well as links between causes and consequences, perception and understanding when it came to the physical world of matter.[15]

This chapter situates the emergence of the novel in the context of the early modern scientific revolution, and the popularization of Newtonian physics throughout the eighteenth century. Public lectures, demonstrations, and printed works intended for men and women of all social classes were important dimensions of the novel's cultural context. In the early part of the century, scientific lecturers such as John Desaguliers and others began to explain Newton's work. They demonstrated to their audience basic features of worldly experience: how the earth's rotation causes day and night, for example, or why things fall down rather than up. How did such discussions of the external physical world conceive of events and their causation? What makes something happen? What makes something stop moving or change its motion? What makes bodies cohere or attract each other? On the one hand, the laws of motion gave indubitable answers to these questions. Newton's laws seemed to be the ultimate triumph of British science, and lent themselves to use as the basis of religious orthodoxy and social conservatism, not to mention military and industrial technology.[16] On the other hand, Newton and his popularizers made it clear that they could not fully explain ultimate causality, though they could and did attribute it to the work of God. In some ways, the more confidently they could explain and mathematically demonstrate patterns of motion in the natural universe, the more remote from our experience the concept of causation became. I want to suggest that such equivocations had an impact on how novels handle causality in their fictional plots. Finally, this chapter will suggest that novelistic plots emerge alongside natural philosophy as a kind of "thinking with objects," as Domenico Bartoloni Meli has called it—or thinking through and around objectivity.[17]

The link between the novel and empiricism is familiar. Perhaps the best-known study of the English novel, Ian Watt's *Rise of the Novel* (1957), begins with a discussion of Descartes, and cites Locke and Thomas Reid, among others. However, Watt portrays the novel's empiricism as a philosophical stance and representational program. He summarizes the novel's empiricism as "the rejection of universals and the emphasis on particulars."[18] Locke's epistemology and views on eloquence are considered; Shaftesbury, Kames, and Hume make brief appearances before Watt turns to social history and the subject of the reading public.[19] But, as the history of science has shown, natural philosophers aimed to produce knowledge about phenomena through a broad range of bodily and epistemological practices and objects. They did more than write about such objects; they actively manipulated them.[20]

If, following Ian Watt and, well before him, Walter Scott, we date the novel's rise from Daniel Defoe (1660–1731), we see that the period of Defoe's life coincides with enormous growth in the success, scope, popularity, and practical application of physical science in Britain, the study of living but especially of nonliving bodies. What appears in eighteenth-century Britain is more like a single culture that encompasses both science and literature, knowledge and entertainment, private and public, and engages men and women (and eventually, children).[21] Throughout the eighteenth century, experimental science became widely popular; scientific instruments such as barometers and telescopes were mass-produced and advertised in almanacs, and scientific lecturers were able to make a living not only in London, but in the provinces.[22] Scientific laws were explained and illustrated in ways that related them to familiar, everyday phenomena and experiments that everyone could perform, or at least envision themselves performing.[23] Such laws, put into simple everyday language, explained the constitution of matter, the movement of bodies, and the effects of force. Concepts such as mass and acceleration figured centrally. Put into concrete practice, they linked the realms of knowledge and of making, including the making of novels.[24]

The homogeneity of material bodies was a central tenet of this philosophy. For example, take the following quotation from *A Course of Experimental Philosophy* (1734), written by John Desaguliers, Newton's protégé and assistant, curator of experiments for the Royal Society, and successful public science lecturer: "That Matter is the same in all Bodies, is evident . . . all matter is homogeneous, or of the same Nature in all Bodies, whether solid or fluid, hard or soft, more or less heavy . . . the whole Variety of Bodies, and the different Changes that happen in them, entirely depend upon the Situation, Distance, Magnitude, Figure, Structure, and Cohesion of the Parts that compound them . . . it follows that there must be a great deal of Vacuity interspers'd in all Bodies . . . and all Spaces are not equally full of Matter. This will be plainly shewed by an Experiment."[25] This passage gives a sense of the scope of natural philosophy's discussion of the external world. Even more importantly, it suggests the homogeneity, unity, and sameness of that world, viewed in a certain way. With a small leap of the imagination, it would be possible to apply this language to a consideration of novelistic plot. Indeed, this preliminary definition of matter in his work, which was intended to dispute what Desaguliers calls Descartes's "philosophical Romance" (n.p.), could even stand as a miniature description of what plot is in a realist novel, what it does, and why: changes in plot happen because of changes in "Situation, Distance . . . Structure, and Cohesion," as well as the spaces between and within bodies.

I will go beyond analogy, however, and suggest that this view of the world reveals a baseline ontology of the fictional medium that a realist novel will shape into a plot. This vision will be complicated over the course of the century in different branches of natural philosophy, as the following chapters will trace, but it will persist at a certain level. I see this assertion of the basic homogeneity of the world, and of the behavior of matter, as the ground of what happens in the plots of realist novels. It underwrites events and causality; it allows an event that happens at one scale to reverberate through diverse realms and scales of experience—it is the ontology that allows the dropping of a window-sash, to return to our initial example, to have effects on the psychological development of a character, or for the bridge of a nose (again, in Sterne's novel) to be linked to another kind of bridge. We could say that it grounds the unity of experience that is relied on both by scientific experiments and by novels' plots. But in practice it does not entail a rigid determinism. The loopholes left by inductive method—the ones that David Hume would go on to enlarge—allow a certain suspense to enter. What appears is an openness of experience, a suspense-filled attention to observing consequences, even if we have predicted what will occur. A plot, like an experiment, shows events as they happen, through the *form* of experience, or through experience considered as a form and test.[26]

Instead of focusing on a single philosopher, I want to situate the novel's emergence within the network of persons, practices, institutions, ideas, and debates that Alexandre Koyré named the "scientific revolution." Far from being a single event, this shift occurred broadly and unevenly over time and across continents. Jonathan Israel's *Radical Enlightenment* explores some of the competing branches and developments of this phenomenon from northern to southern Europe, England, Russia, and Scandinavia. Discussing the influential late seventeenth-century learned periodicals that flourished across Europe in multiple languages (of which the *Philosophical Transactions of the Royal Society* was one example), Israel notes their "unceasing advocacy of . . . unbiased detachment and deferring judgment" in the face of proliferating data.[27]

Steven Shapin has recently argued on behalf of resurrecting the concept of the scientific revolution (with appropriate scruples) in light of the early modern period's "self-conscious and large-scale attempts to change belief, and ways of securing belief, about the natural world."[28] This framework is useful for understanding the novel because of how it opens up and extends knowledge of the inanimate world through particular forms of practice and experience. Natural philosophy in this period attained a new level of mathematical certainty. But it also backed away from questions of ultimate causes in ways that perhaps enabled the refinement of long-form narrative fictionality.

A core issue for investigation was motion as it occurs in the physical world: how and why inanimate things move the way they do, and at what level or scale movement should be explained. Laws and models challenged the received ideas of Aristotelian physics. The latter, as Adam Smith explains in his posthumously published *Essays on Philosophical Subjects* (1795), held that "the natural motion of the Earth . . . was downwards . . . fire and air was upwards. . . . Each of the Four Elements having a particular region allotted to it, had a place of rest, to which it naturally tended, by its motion, either up or down, in a straight line, and where, when it had arrived, it naturally ceased to move."[29] In the peripatetic philosophy, there was, in addition to natural motion, "violent motion": "Natural motion was that which flowed from an innate tendency in the body, as when a stone fell downwards: Violent motion, that which arose from external force, and which was, in some measure, contrary to the natural tendency of the body, as when a stone was thrown upwards, or horizontally. No violent motion could be lasting . . . it would soon be destroyed."[30] These ideas of motion, still widely taught and officially sanctioned until the mid-seventeenth century, relied on an animist and teleological view of the movement of matter, in which inanimate things "aspire to be at their natural place," and thus seem to possess "purpose, intention, or sentience."[31] In Aristotelian thought, in Jonathan Israel's words, "Behaviour and function . . . are determined by, the soul or essence of things rather than mechanistically."[32] Things, in other words, possessed distinct characters.

Mechanical philosophy challenged this view. It asserted that things had their physical being not as substantial forms, as in Aristotelian philosophy, but only as compositions or contextures of small particles or bodies that were subject to but distinct from the dynamic forces that moved them. Matter could be regarded as homogeneous, as seen in the phenomenon of objects falling at uniform velocity. The canonical air-pump experiment involving falling bodies is cited by nearly every writer. Here is a typical example from an

Introduction to Newton's Philosophy written by the influential Willem s'Gravesande, who would go on to become professor of physics at Leiden: "Bodies which descend by the force of gravity . . . fall with the same velocity. Which is proved by an Experiment.] Pump out the air from the tall recipient AB . . . let fall a peice [*sic*] of gold and a very light feather at the same time, and they will always come down . . . at the same instant of time."[33] Robert Boyle explained with great clarity the "corpuscular or mechanical philosophy" as resting on only two principles: "Matter and motion. . . . There cannot be any physical principles more simple than matter and motion; neither of them being resoluble into any other thing."[34] The tiny bodies of matter, Boyle argued, have either a geometrical shape or an irregular one, and have particular ways of moving: "What a vast number of variations may we suppose capable of being produced by the compositions, and recompositions of myriads of single invisible corpuscles. . . . And the aggregate of those corpuscles may be further diversified by the texture resulting from their convention into a body."[35]

This philosophy claimed to be simple and intelligible. Boyle insisted that "men do so easily understand one another's meaning, when they talk of local motion, rest, magnitude, shape, order, situation, and contexture, of material substances."[36] Desaguliers strikingly illustrates the idea that all things in the physical world depend simply on the way that basic particles of matter are arranged:

> The same Atoms being as proper to make Land as Sea, to make Gold as Clay. . . . One may bring various Examples of Matter trac'd through several Bodies, whose Changes depend upon the different Texture and Position of the Parts. . . . If we had consider'd such Drops of Rain as impregnated Linseed, we might have trac'd it thro' the Stalks of the Plant, Flax made of that, Thread spun from the Flax, Linnen made of the Thread, a white Pulp made of the Rags of the worn Linnen beaten up with Water at the Paper-Mills, Paper made of that Pulp thinly spread upon a fine Net-Work of Wire; and lastly the Smoke, which burn'd Paper affords, is again easily reduc'd to Water.[37]

The same bodies continually rearrange themselves to assume what appear to be successive identities. (Notice that in this example, the same atoms constitute the plot of the story and the material pages on which it is printed and read.[38])

The phenomenon of motion became the center of attention in the mechanical philosophy. Its partner, matter, is that which is moved, or changes its place. Its passivity, or in Adam Smith's words, "the real inertness of matter," was insisted on repeatedly, though vitalist philosophies disputed this picture of the world.[39] Beginning in 1692, the annual lectures endowed by Robert Boyle, an institution that played an important role in propagating Newtonian ideas, tried to disprove the notion that matter could move itself: "If matter moved by its own inherent force, God would be rendered useless."[40] It was important to refute Spinoza's view that motion was inherent to matter and that mechanistic cause and effect applied to moral and spiritual phenomena; Jonathan Israel traces the long battle between what he calls the "moderate mainstream" of Newtonian thought and a "radical fringe" of Spinozist materialism that goes underground rather than disappearing.[41] Popular explanations of Newton's thought, such as we find in Desaguliers or in Oliver Goldsmith's *Survey of Experimental Philosophy*, define matter in terms of its inertness. In Goldsmith's account of Newtonian laws, "Matter is of itself entirely passive, incapable of moving itself, or stopping its own motion; a ball thrown by the hand would continue to go on for ever, did not the force of attraction, or the resistance of the air, at last destroy the motion it received from the slinger's arm. Matter, in short, follows whatever direction is impressed upon it, and is affected by every impulse in its way."[42] Or again, in Smith's words, "The real inertness of matter" lies "not in an aversion to motion, or in a propensity to rest, but in a power of continuing indifferently either at rest or in motion, and of resisting, with a certain force, whatever endeavoured to change its state from the one to the other."[43]

But then what makes material things move? As such quotations suggest, both of which paraphrase the first law, Newton's laws of motion were often understood not only through the widely read Boyle lectures, but through demonstrations and publications that aimed to reach all types of audiences. As early as 1713, Desaguliers notes, "I have with great Pleasure seen the Newtonian Philosophy so generally received among Persons of all Ranks and Professions, and even the Ladies, by help of Experiments."[44] Benjamin Martin, who made his living as a provincial science lecturer before settling into a career as an instrument-maker in London, published in a one-shilling edition, *A plain and familiar introduction to the Newtonian philosophy, in six sections illustrated by six copper plates. Designed for the use of such gentlemen and ladies as would acquire a competent knowledge of this science, without mathematical learning* (1754). Rather than taking it for granted that things moved because they in some sense wanted to, Newtonian philosophy made motion, its occurrence and trajectories, into complex problems and practical challenges, with potentially large uses.[45]

I want to think about the implications of this philosophy for plot, for how things happen.[46] It became possible to think of motion as possessing something like a narrative form. This is suggested by the way Newton lays them out verbally, and the scenarios he invokes to explain them. "Every body perseveres in its state of rest, or of uniform motion in a right line, *unless it is compelled to change that state by forces impressed thereon*," the first law states (emphasis added).[47] The second axiom: "The alteration of motion is ever proportional to the motive force impressed; and is made in the direction of the right line in which that force is impressed." Finally, the third is thus glossed by Newton himself in a way that makes it not only familiar from experience but almost plot-like:

> Whatever draws or presses is as much drawn or pressed by that other. . . . If a horse draws a stone tied to a rope, the horse . . . will be equally drawn back towards the stone: for the distended rope, by the same endeavour to relax or unbend itself, will draw the horse as much towards the stone, as it does the stone towards the horse, and will obstruct the progress of the one as much as it advances that of the other. If a body impinge upon another, and by its force change the motion of the other, that body also . . . will undergo an equal change, in its own motion, towards the contrary part.[48]

Benjamin Martin helpfully points out that "on this principle it is, that fishes swim, and birds fly, and men row a boat, because when the medium is acted upon by the fin, wing, or oar, in any one direction, by an equal reaction, it moves the fish, the bird, and the boat in a direction just the contrary."[49]

Newton's laws account for what he calls "the frame of the System of the World."[50] He explains how it is that moons and planets do not fly off in straight lines, or fall in on each other. Adam Smith describes this system as "an immense chain of the most important and sublime truths, all closely connected together, by one capital fact, of the reality of which we have daily experience."[51] Desaguliers notes, "It would be endless to shew all the Consequences of the two Laws of Motion already explain'd, in the Practice of mechanical operations; and to apply them to the Explication of all kinds of Motions whether of Bodies on Earth, or of the Planets and Comets in the Heavens."[52] While the triumph of Newton's physics obviously derived (at least until the twentieth century) from their accuracy and applicability, these axioms, as popularized in simple form, also add a potentially narrative or plot-like character to the world's phenomena. Some read almost like extremely abstract plot summaries. A character, let's say one named Clarissa, wishes to keep up her "uniform motion in a right line," but is "compelled to change that state" by external forces. Will that body that impinges on her not undergo an equal change?

But I want to suggest that a dramatic or plot-like quality can also pervade small or local motions, even when their endpoint or origin is not apparent. The way things are moving, how fast or slow, in what direction they move: none of these things can be taken for granted; nor can force, on which a new kind of attention is brought to bear. If a body changes its direction of motion, if it does not move in a straight line, or if it speeds up or slows down, that means there was some "motive force impressed," which involves the existence of another body. If force is applied to a body, it changes its way of moving in that same line. In Newton's gloss on the second law: "If any force generates a motion, a double force will generate double the motion, a triple force triple the motion, whether that force be impressed altogether or at once, or gradually and successively." Motions and forces can be measured, added together or subtracted to "produce a new motion compounded from the determination of both."[53] Time becomes a critical dimension of thought, motion becomes a phenomenon that calls for explanation, and inertia (whether rest or motion) becomes a mode of being, a part of a story.[54] The action of force over an increment of time can be measured, predicted, and applied. But it also becomes an event with a duration, a before and an after. The mutuality of effect announced in the third axiom, for example, seems particularly rich in narrative implications: something that effects a change in motion will itself be equally changed.

The homogeneous nature of the physical world is another Newtonian axiom on which realist plots will rely.[55] In his "Rules of reasoning in philosophy," with which Newton begins the third book, he writes,

> Rule I. We are to admit no more causes of natural things than such as are both true and sufficient to explain their appearances . . . more is in vain when less will serve; for Nature is pleased with simplicity, and affects not the pomp of superfluous causes.
>
> Rule II. Therefore to the same natural effects we must, as far as possible, assign the same causes. As to respiration in a man and in a beast; the descent of stones in Europe and in America; the light of our

> culinary fire and of the sun; the reflection of light in the earth and in the planets.[56]

Matter behaves in the same way and possesses the same characteristics; the light from the kitchen fire is the same as that which comes from the sun. This is what allows Newton to move from examples like a stone being swung around in a sling, or a suspended twirling bucket full of water to the system of the universe. But if it can familiarize the moons of Jupiter, it can also defamiliarize local terrestrial experience.

Newton invokes other possible worlds in ways that can be quite striking. In his definition of centripetal force, for example, as "that by which bodies are drawn or impelled, or any way tend, towards a point as to a centre," he points out that were it not for the centripetal force of gravity, "A projectile . . . would go off from [the earth] in a right line . . . nor could the moon without some such force, be retained in its orbit. If this force was too small, it would not sufficiently turn the moon out of a rectilinear course; if it was too great, it would turn it too much, and draw down the moon from its orbit towards the earth."[57] The fact that things move in stable, regular orbits around another body becomes a phenomenon in need of explanation and mathematical description. As understanding increases, so might a sense of wonder at this system. Toward the end of Book 3, Newton remarks that "this most beautiful system of the sun, planets, and comets, could only proceed from the counsel and dominion of an intelligent and powerful Being. . . . lest the systems of the fixed stars should, by their gravity, fall on each other mutually, he hath placed those systems at immense distances one from another."[58] In this we could see what Jonathan Israel calls the "new form of triangular partnership between science, philosophy, and theology."[59] Contemporaries and followers such as Richard Bentley and Samuel Clarke availed themselves of such observations in order to support orthodox Christianity and a certain vision of social and even economic relations.[60] But the writings of Newton and some of his other followers do not hesitate to point out their inability to answer the question of who or what might be responsible for the features of such a system, of how it came into being.

Newton lays out "the laws and conditions of certain motions"; they describe the necessary behavior in time and space of certain arrangements of matter.[61] But he ends the *Principia* with the famous reflection that

> hitherto I have not been able to discover the cause of those properties of gravity from phaenomena, and I frame no hypotheses. . . . In

> [natural] philosophy particular propositions are inferred from the phaenomena, and afterwards rendered general by induction. Thus it was that the impenetrability, the mobility, and the impulsive force of bodies, and the laws of motion and of gravitation, were discovered. And to us it is enough that gravity does really exist, and act according to the laws which we have explained, and abundantly serves to account for all the motions of the celestial bodies, and of our sea.[62]

s'Gravesande and Desaguliers are likewise explicit about the limits of the knowledge produced by natural philosophy: "We call a Law of Nature," s'Gravesande begins his *Introduction to Newton's Philosophy*, "every Effect which in all Occasions is produced after the same manner; although its Cause is unknown to us."[63] Of attraction and repulsion, he concludes that "the Phaenomena of Nature shew that there are really such Principles, tho' it has not been yet explain'd what their Causes are"; he further notes, "From two or three Phaenomena of Nature to deduce general Principles of Motion, and then explain in what manner the Properties and Actions of all Things follow from those Principles, would be a great Progress made in Philosophy, tho' the Causes of those Principles should not yet be known."[64] Desaguliers cautions his readers: "N.B. When we use the Words Gravity, Gravitation, or Attraction; we have a Regard not to the Cause, but to the Effect; namely to that Force, which Bodies have when they are carried towards each other . . . whether it be occasion'd by the Impulsion of any subtile Fluid, or by any unknown and unmechanical Power concomitant to all Matter" (6–7).[65] Even Fontenelle's celebratory *Life of Sir Isaac Newton* points out, "It is not known wherein the nature of Gravitation consists, nor was Sir Isaac Newton himself acquainted with it. . . . He makes use only of this Word upon all occasions to express the active Force of Bodies; a Force indeed unknown, and which he does not pretend to define."[66]

Evander Agazzi sees this "anti-essentialist and anti-substantialist" stance as the hallmark of modern science, which he dates to Galileo. It is not concerned with the ultimate causes of empirical phenomena, but with the mathematical description of their behavior or "affections."[67] Steven Shapin perceives this attitude behind the modern turn toward the production of "matters of fact" through new institutions and procedures.[68] Such facts depended on certain conventions, both social and epistemological. Because causes can only be inferred from observable effects, they constitute a probabilistic form of knowledge and merit at most a conditional and limited assent.[69] This attitude proved to be robust and long-lasting. In 1881, John Stuart Mill wrote, "I make no research into the ultimate or ontological causes of anything. . . . The only notion of cause which the theory of induction requires, is such a notion as can be gained from experience."[70]

But the desire for causal explanation did not disappear once accurate calculations of movement became available. Rather, it took a particular turn, and this is where we might see a new opening for the kinds of plotting that novels perform. Even if ultimate agency within the Newtonian world was still referred to God, as Courtney Weiss Smith has argued, there was still room to explore in detail the work of *intermediating causes*.[71] s'Gravesande thus defines "natural phaenomena," the proper object of natural philosophy: "Natural phaenomena are all situations and all motions of natural bodies, not immediately depending upon the action of an intelligent being; and which may be observed by our senses. . . . We do not exclude out of the number of natural phaenomena, those which happen in our bodies by our will; for they are produced by the motion of our muscles, and their action depends upon another motion: in these, there is only that motion which arises from the immediate action of the mind, and is entirely unknown to us, which is not a natural phaenomenon."[72] As long as there is an observable sequence of motions, one bit of matter acting on another bit of matter, there is no need to understand that initial push or impulse. s'Gravesande goes on to note, "We call a law of nature, every effect which in all occasions is produced after the same manner; although its cause is unknown to us"; by "cause," he means how exactly these movements are connected to "the will of god" or to a movement of the mind.[73]

In Boyle's account, the mechanical philosophy concerns itself only with these intermediating causes that produce the things or effects that we see. In a way that anticipates the arguments of actor-network theory, Boyle rejects explanations that point to agency as something transcending empirically observable effects: "When, to solve the phenomena of nature, agents are made use of which . . . are such that we conceive not how they operate to produce effects; such agents I mean, as the soul of the world, the universal spirit, the plastic power etc., the curiosity of an inquisitive person is not satisfied hereby; who seeks *not so much to know what is the general agent that produces a phenomenon, as by what means, and after what manner, it is produced*."[74] What philosophers should seek, according to Boyle, is not some "general agent" but the chain of minute, physical, material things that add up to produce an effect: "Not so much what the agent is or does as, what changes are made in the patient . . . by what means, and after what manner, those changes are effected."[75] Boyle appears to distinguish between a metaphysical idea of agency and a more limited, physical one that's understood to be intermediate. He points out that identical changes can be "effected by the same, or the like means; or after the same, or the like manner" regardless of the agency involved: "If corn be reduced to meal, the materials and shape of the mill-stones and their peculiar motion and adaptation will be much of the same kind; and . . . the grains of corn will suffer a various attrition . . . whether the corn be ground by a water-mill, or a windmill, a horsemill, or a handmill; that is, a mill, whose stones are turned by inanimate, by brute, or by rational agents."[76] If an angel wanted to grind corn, Boyle goes on to say, they would still need to find something hard to do it with. Material effects require material causes, and those causes are defined as intermediating ones. They come in between some mysterious impulse, volition, or agency on the one end, and the physical change in the world that we can perceive on the other end of the sequence. Boyle's term is "local motion": the way that the parts of things are put together to push, pull, and counterbalance each other. Arguing against the Aristotelian doctrine of substantial forms, Boyle claims that "*things are effects*; so that, by whatever principles natural things are constituted, 'tis by the mechanical principles that their phenomena must be clearly explained."[77] Things are effects, maybe even stories. So the question to ask is not, "What is it?" but "How does it work, what things, put together in what sequence, make it move and produce effects in the external world?" As we will see, novels such as Sterne's *Tristram Shandy* will pose these kinds of questions.

The notion of intermediating causes could still appear mysterious, however. Goldsmith remarks that natural philosophy can only study "that motion, which is communicated from one body to another, without considering the first cause which gave that motion to either."[78] Nor can even that intermediate process be fully understood: we cannot really know "how one body becomes possessed of a power of granting its motion to another." It can, however, be measured as a "precise quantity."[79] And as long as it can be observed, measured, and manipulated, this chain of intermediating causes can be put to use in the physical world:

> It appears by Observation that Gravity is the Cause of the Fall of heavy Bodies, which observe certain Laws in their Motion—That a heavy Body by its Descent moves the Axis of a Wheel, that carries round another by its Teeth, which by the Intermediation of other Wheels

> and Pinions, carries round a Hand upon a Dial-Plate to measure Time, or for other Uses; it wou'd be very unphilosophical to say—That our Reasoning about the Cause of the Motion of the Hand is false, as being founded upon occult Qualities; because we can go no higher than Gravity, whose Cause we don't pretend to know.[80]

The focus shifts to those interlocking mechanisms, then, that carry out a movement or action, that provide the links between a mysterious beginning and a definite end. These are causes that are intermediate in space and time and flesh out these abstract concepts. The plots of realist novels do something similar.

An important means of investigating such intermediating causes and effects was what Peter Dear has called the event-experiment: "The new scientific experience of the seventeenth century was characterized by the singular, historical event experiment, which acted as a surrogate for universal experience."[81] Aristotelian science had little use for singular events or specific experiences: the experience that counted was knowledge of "how nature usually behaved; it did not consist of knowledge of discrete events, because such events might be anomalous."[82] But innovative experiments proliferated in the seventeenth century: "The Torricellian experiment, the moons of Jupiter, sun spots, the circulation of the blood, the differential refrangibility of light—and this novelty generated problems for the establishment of knowledge."[83] Dear traces the historical transition from the requirement of supposedly universal, nonspecific experience, to the acceptance of the singular event: a scientist like Niccoló Cabeo, for example, investigating the speed of falling bodies, remarked, "Both I myself and others have tried this by many experiments, and I have always observed [the weights] to fall in an exactly equal time," before folding these particular instances into a general statement, supported by the expertise he is claiming for himself: "Cabeo . . . avoids narrating specific trials, framing his claims as already universalized statements of experience: 'they reach the ground' . . . and so on. The scientific 'experience' subsumes the frequently repeated individual instances in the usual Aristotelian fashion."[84] In the case of the Jesuit astronomer, Giovanni Battista Riccioli, instructions for performing experiments come to play a crucial role. Telling the reader how to set things up and then describing the result, Riccioli, in Dear's words, "leaves unquestioned the truth of the universal experience that the instructions purportedly enable the reader to realize; in effect, the instructions act as an elaborate statement of the 'proposition' itself."[85] Galileo also stands midway between the Aristotelian notion of universal experience and the specific event-experiment; he refers to numerous repeated trials that he has made, but doesn't narrate exactly what happened on particular occasions, referring instead to a generalized outcome: "He told his reader *what happens*."[86] It is in this context that we need to see the Royal Society's elevation of the event-experiment. Instead of offering a "universal statement of how some aspect of the world behaves," the typical Royal Society report told "how, on one specific occasion, the world had *behaved*. . . . The crucial element in these reports is the rhetorical establishment of the actuality of a discrete event . . . [they] stress the particularities of the singular instance itself, as if the establishment of belief in the historical event was itself sufficient."[87] In other words, believing that the event described had happened one time came to warrant believing that it would happen that way every time. In the familiar Aristotelian opposition, history relates what happened one time to a particular person, while poetry relates the kind of thing that necessarily or probably happens. The event-experiment offers another alternative, one quite relevant to the way things happen in realist novels. Because of the way it's embedded in a particular set of circumstances, the event-experiment becomes a universally repeatable one, at least in theory.

We can see a shift in the concept of experience, and a different profile or structure attaching to the notion of an event. The universal experience of a phenomenon, such as bodies falling, did not make redundant a singular event-experience, such as a carefully arranged experiment inside an evacuated air-pump receiver. The latter kind of experience became something that could command belief or fail to command belief. If it did succeed in achieving credibility (something that depends, as Shapin, Schaffer, and others emphasized, on social conventions and status), that belief had to cover or encompass a series or sequence of microevents, closely observed, that followed one another consecutively. This can be seen in the way that s'Gravesande and others give instructions for experiments, for example, showing capillary action: "Experiment 3.] Immerge in Water the Ends of small glass Tubes open at both Ends. . . . The Water will spontaneously ascend in them, and so much the higher as the Diameter is less."[88] Here are Goldsmith's instructions for a familiar experiment in magnetic attraction: "Strew some steel filings over a sheet of white paper, and underneath, let a magnet be placed; then gently tap the paper with the finger . . . so as to give the filings a little motion, and they will soon be seen to form a regular striated figure upon the paper."[89] All of these experiments became familiar through public lectures, demonstrations, and publications;

they're also narrated here to give the sense of what Shapin and Schaffer call virtual witnessing.[90] In the following passage, Goldsmith collapses the account of an event that happened in the past, a hypothetical situation, and a set of minutely detailed predictions about a sequence of microevents.

> It was . . . soon after perceived, that if the touching non-electric body, was placed in such a manner as to touch nothing but the electrified globe or tube, and to have no communication with any other body, it then became electric itself. . . . If, for instance, a piece of metal . . . was fixed on the top of a glass tube . . . the tube, being electrified, by rubbing with the hand, the metal above was seen itself to become electric; and . . . when the finger was brought near it, about the distance of half an inch, to snap and crackle. . . . Let us now then suppose the whole electrical apparatus prepared, the glass globe turning swiftly upon its axle . . . while the person to be electrified stands upon a cake of resin, wax, and sulphur, mixed together, of about fifteen inches in diameter, while he touches the upper part of the globe. . . . The weather being dry and the room spacious, we shall see the following wonders ensue. In a few seconds, the man will be filled with effluvia; he will become perfectly electrical; his hands, and every part of his body will attract and repell light substances at four feet distance. . . . Each of these little substances, such as straws, motes, and leaf gold, are at first drawn towards the electrified body with great swiftness, and then . . . repelled with equal force.[91]

This account seems to rest somewhere between a singular event and universal experience, between prescription, description, and fiction. It's not clear whether he's describing something he witnessed one time or many times; presumably, it is something that will happen in any future moment if the conditions are faithfully enacted. Likewise, the cause is unexplained. Goldsmith explicitly calls "attraction, or that power by which we see one body approach another," including magnetism and electricity, "a secret that human sagacity has not yet discovered; we are certain of the fact, we see plainly that the substances do approach, and all that we can assign as the cause of their coming together is but conjecture . . . yet the experiments that serve to prove it are incontestable."[92] The sequential quality of the narrative draws our attention, generates suspense, commands belief in its own right. The man stands, the glass globe whirls around, he puts his hand on it, and we wait to see what happens. Equally vivid is the imagined registration of what ensues, the way our senses register the "following wonders."[93] Finally, the conditions or circumstances, what we could normally call the setting, cannot be separated from the events that occur: the set-up and the event confirm each other's validity. If the effect described is produced, that confirms that the set-up was done correctly, which in turn legitimates the event as more than a one-off happening or accident. In conjunction with its setting, the event comes to possess an illustrative significance.

The senses have a vital role to play: they testify to the objective existence of these movements or effects, and their vivid testimony compensates, as it were, for the lack of complete knowledge of the causes behind these physical phenomena. But they also have to put into place something like negative capability, or an ability to refrain from overhypothesizing. In Desaguliers's words, "Those Properties of Bodies, such as Gravity, Attractions, and Repulsions . . . do really exist, and are by Experiments and Observations made the Objects of our Senses. These properties produce Effects, according to settled Laws, always acting in the same Manner under the same Circumstances: And, tho' the Causes of those Causes are not known, since *we do not reason about these hidden Causes*; it is plain that we reject occult Qualities, instead of admitting them in our Philosophy, as the Cartesians always object to us."[94] In another twist, one of the most significant features of Newton's natural philosophy was the way that some of its discoveries appeared to contradict everyday sense experience, and this again suggests that realism's registration of experience is far from simple. As Helen Thompson argues, "Eighteenth-century novels make explicit the *production* of empirical reality as the reader's encounter with forms and powers that enable sensational knowledge."[95] But sometimes readers have to draw back from their own encounters with the world. Some sensations have to be questioned, canceled, or subordinated to abstract forms. The first law of motion, for instance, requires some intensely counterfactual imagining, as Desaguliers clearly acknowledges. The idea that an external impulse is needed to set a body in motion is not difficult to grasp, he notes,

> But we don't so readily perceive that a body in Motion would continue to move for ever without the action of an extrinsical agent; because we see Bodies here on earth gradually lose their Motions, and for want of attending to all the Causes that destroy the Motion of Bodies, we often imagine that Motion languishes and at last quite perishes of itself. But if we consider what external Causes retard and destroy

> Motion, we shall soon perceive that if those Causes were remov'd, a Body once put into Motion in any Direction would continue in that Motion and Direction for ever.
>
> A Stone thrown forwards with the Hand goes on with Motion that it has received from the Hand, and would continue in that State of Motion for ever, if there was neither Air nor Gravity.[96]

Here, instead of registering a phenomenon through the senses, you are asked to imagine a world without air or gravity in which you would throw a stone and watch it go on forever. Desaguliers actually asserts that we can "perceive" that phenomenon, if we manage to accomplish a peculiar exercise. We have to imagine the operation of "external Causes" on a body in motion, and then negate or remove the world around it.

The unreliability of the senses is taken up in another work of popular Newtonian philosophy, *Astronomical Dialogues Between a Gentleman and Lady; Wherein* [. . .] *the Elements of Astronomy and Geography are Explain'd, In a Pleasant, Easy, and Familiar Way* (1719), written by John Harris.[97] When "Lady M" expresses her desire to "know something of the first Principles and Rudiments of Astronomy," the narrator explains how to use and understand the terrestrial and celestial globes in her drawing room.[98] In between the exchange of poetry and fulsome compliments, Lady M points out the discrepancies between science and sense experience. They don't feel the earth spinning, for example. "I don't know any thing that fastens us down to the Earth," she remarks.[99] The narrator obligingly explains "the Law of Gravity, or Gravitation," and refers to the phenomenon that would later come to be called vection, or the inability to tell whether you're in motion or the outside world is moving. His example is of two ships: "Let the Vessel move never so fast forwards, if it were not for the Tossings and Shocks which the Resistance of the Water and Waves make . . . you would perceive no Motion at all in the ship, but judge it to be perfectly at Rest; and if another Ship lay at Anchor by you, you would judge that to move *backwards*, and not your self *forwards*."[100]

Others stress this phenomenon as well. Benjamin Martin observes that "*if the eye be in motion*, an object before it, tho' at rest, will have its image move over the retina, and so produce a *deceptive idea of motion*."[101] Distinguishing between "real and apparent motion," Martin points out that "we may not only have an idea of motion in bodies when they do really move, but also when they *do not move*, but are absolutely at rest. This ambiguity in our ideas of motion renders us subject to many and very great mistakes in our common notions or judgment of things."[102] Using the same example as Harris, he notes that inside the cabin of a ship in motion, you can't feel that you're moving at all. If another ship moves by you, "If you were in A [moving at a slower velocity], the ship B would appear to go forwards from you . . . if you were in B, the ship A would appear to go backwards or the contrary way"; he warns that judgments about motion that rely on the senses are "precarious, and mostly full of illusion and error."[103] Yet Harris and Martin do not seem too worried about this built-in "ambiguity in our ideas of motion." They assume that this phenomenon can just be pointed out. This body I see may or may not be moving in reality, but I know that there is a regular causal framework that legitimates this curious experience of suspended belief that is required of me.

Natural philosophers asserted that intermediating causes in the physical world were knowable, measurable, and manipulable, in productive and limited ways. This knowledge didn't encompass first causes or the natures of things, and yet it could have the effect of undercutting sense experience or, to the contrary, weighting it with unusual significance. I want to suggest that this provides an important context for the development of fictionality in the novel: for the ways in which novels develop their elaborate, thing-filled, motion-filled plots, full of effects whose causes often have to be assumed, and events that depend on a precise set of circumstances. Novels also play with abstractions. We can find a suggestive analogy for what novels do in the phenomenon that Domenico Bertoloni Meli has called "thinking with objects." I will argue throughout this book that novels' plots depend on varied sensory inputs and ways of engaging with material phenomena—for example, through handling, pushing, being moved, dissolving. Bertoloni Meli argues that the science of mechanics developed through the ways that early modern mathematicians, engineers, and others interacted with concrete objects: the pendulum, pierced cistern, strings, levers, inclined planes, and other simple machines, in addition to newly invented scientific instruments. Stressing the "polyvalence in the notion of objects" that prevailed in this period, he suggests a role that could only be filled by the senses and by tangible things: the pendulum, for example, functioned "both as a conceptual tool to explore oscillations and vibrations . . . and as an experimental tool to investigate collision."[104] Objects were tangible, concrete, but also remained available as abstractions. In his book on mechanics, for example, the Renaissance scholar Guidobaldo dal Monte offered side-by-side illustrations of actual machines and geometrical diagrams of them. Mathematicians were aware of the differences between material objects and their "abstract counterparts: perfectly smooth inclined planes, perfectly elastic collisions." Bertoloni Meli argues that the knowledge created by these mathematicians depended on their manipulation of concrete experimental objects, and their practical knowledge of how such things behaved; but the final product, laws and principles of motion, "aimed at abstracting from the peculiarities and vagaries of matter."[105]

We can perhaps think of realist novels as performing this operation in reverse through their plots. They resubmerge abstract knowledge about how things move and behave in "the peculiarities and vagaries of matter"; they place laws and formulas back into concrete situations and contingencies. For novels, the knowledge in question also concerns intermediating causes and the ambiguity of sense-perception, which is at once necessary and often misleading. Events, causes, and circumstances are also polyvalent in novels. They exist as particular things, tied to unique individuals, and yet they also persist as abstract configurations of force and matter. In the rest of this chapter, I will take Sterne's *Tristram Shandy* as an example of how a novel can think and feel with objects. It shapes its fictionality and builds its plot around the behavior of matter in the physical world, which is taken to be both abstract and concrete, factual and counter to fact.

My focus on this particular novel may seem to make things too easy. The pages of Sterne's novel are filled with machines and mechanical movements, real or metaphorical.[106] *Tristram Shandy* even refers to itself as a machine of a particularly complex and intricate type, as critics have remarked.[107] Sterne's work may appear too eccentric to count as a novel, much less a realist novel.[108] Yet I will suggest that it is indeed the most typical of novels in its commitment to thinking with objects and in terms of intermediate causes. Shklovsky states that he will use *Tristram Shandy* "to illustrate general laws of plot"; he thus gives it a status analogous to something like the maximum-machine, a device that was used in the eighteenth century to demonstrate how much force could be exerted over time by one body.[109] I will look at how Sterne's plot incorporates and, in a sense, gives in to the forces and matter of the physical world. *Tristram Shandy*, as every reader has noticed, does not have a conventional plot structure, though it certainly obeys what E. M. Forster calls "the clock," or the embedding of referential time within the novel (indeed, Sterne gives this clock more prominence than most authors, in his attention to the time of narration vs. the time of events, and of course begins his novel famously with the winding of a clock—the kind of winding-up that played an important role in debates about the religious implications of Newton's system).[110] While much attention has gone to the subject of the digressions in his narrative (which Sterne eye-catchingly diagrams), I want to focus on sequences. Sterne could be said to replace plot as synthesis with plot as intercalated sequence.

But these sequences are related not only in time: *Tristram Shandy* is keenly aware of sequences of motion that are carried out (or just as often interrupted) by the direct contact or pressure of one body against another: the chains of intermediating causes that Boyle and others highlight as the proper subject of investigation. For example, here is how the novel explains the cause of Uncle Toby's sudden departure for the country, while convalescing from his wound. Toby had been studying military architecture, which led him to the "sweet fountain of science," the projectile and motion studies of "Gallileo and Torricellius," which he then abandons to return to "the practical part of fortification only; the pleasure of which, like a spring held back, returned upon him with redoubled force" (79–81). One day, at his small table, he "had the accident, in reaching over for his tobacco-box, to throw down his compasses, and in stopping to take the compasses up with his sleeve he threw down his case of instruments"; next, "In his endeavouring to catch the snuffers in falling, he thrust Monsieur Blondel off the table, and Count de Pagan o'top of him" (83–84). This series of movements leads his servant, Trim, to conceive of the project that becomes Toby's famous "hobby-horse": the building of miniature scale-model towns and fortifications on the bowling green of his house, complete with "batteries, saps, ditches, and pallisadoes" (86). Trim quickly proposes using these to reenact the campaigns of the Nine Years' War, starting "on the very day that his Majesty and the Allies take the field, and demolish 'em town by town as fast as—Trim, quoth my uncle Toby, say no more" (86–87).

That chapter illustrates two important features: attention to a sequence of physical microevents, and a type of "thinking with objects." The physical manipulation of objects, including human bodies, and even the attribution of human qualities to them (agency, feeling, etc.) turns out to be this novel's best way of finding meaning in necessarily truncated sequences of causality. First causes are not empirically verifiable, while finality, Sterne suggests, is not representable. We are left with the way that material events are linked together in the middle of a sequence. Yet that sequence may be interrupted by the insertion of another sequence, as we see above, when Toby tries to stop one object from falling, and in the process sets into motion another sequence in which one book falls and causes another to fall, and so on.

The type of thinking with objects that we see in *Tristram Shandy* involves a lively cross-referencing of material thing and abstract or geometrical shape;

for instance, Walter Shandy reaches with his left hand for the handkerchief in his right pocket: "When my uncle Toby discovered the transverse zig-zaggery of my father's approaches towards [the pocket], it instantly brought into his mind those he had done duty in, before the gate of St. Nicholas; . . . the returning angles of the traverses of that attack—but particularly of that one, where he received his wound upon his groin" (144). Here the thought goes from body to geometric shape (Toby almost sends for his compasses to measure the angles) and back to his own body. The most elaborate instance of thinking (and feeling) with objects can be found, of course, in Toby's scale-model fortifications, which result in the production not so much of scientific knowledge as of embodied historical experience. As Toby receives the latest news of the battles being fought during the War of the Spanish Succession, he takes the map of the besieged town, and then meticulously "by means of a large role [*sic*] of packthread, and a number of small piquets driven into the ground, at the several angles . . . he transferred the lines from his paper" (400).[111] He adds another dimension to his "little plot" through miniature ditches and embankments (and later on, further realistic embellishments), and then reenacts the events, "regulating their approaches and attacks . . . step by step with the allies" (401). Paradoxically, by becoming abstract, the movements of the army become part of his bodily experience, and "hurried on the blood" (419).

Thinking emerges from objects but then appears to enter back into them, giving them an appearance of animation or even intelligence.[112] Corporal Trim, for instance, announces the death of Tristram's older brother and drops his hat on the floor in a dramatic gesture: "Twas infinitely striking! Susannah burst into a flood of tears. . . . The descent of the hat was as if a heavy lump of clay had been kneaded into the crown of it.—Nothing could have expressed the sentiment of mortality, of which it was the type and fore-runner, like it" (325–26). The movements of bodies, likewise, seem to mediate ideas and feelings in an equally mysterious way. Tristram's father, on receiving bad news, responds the first time by walking upstairs and throwing himself on his bed, but the second time by going quietly out to the fishpond "with the gentlest movement of limbs." The narrator comments that in such cases, nature "determines us to a sally of this or that members . . . or posture of body, we know not why . . . even the clearest and most exalted understandings amongst us find ourselves puzzled and at a loss in almost every cranny of nature's works" (263). Causes remain inaccessible—even, sometimes, the causes of why things or people fall down. Indeed, the more detailed the novel's description of any motion, the more elusive its cause seems to become.

Because of this imperfect knowledge, thoughts seem to be half objects, and objects appear to be half made of thoughts. But both thoughts and things figure in *Tristram Shandy*'s plot sequences as subject to the forces of pressure, friction, gravity, and velocity. These phenomena furnish the miniature plots of the sequences that have the strongest internal cohesion or identity (even when interrupted), and underline the common situations and motions shared by thoughts, feelings, animate and inanimate things. The narrator notes Walter Shandy's belief that "knowledge, like matter . . . was divisible in infinitum;—that the grains and scruples were as much a part of it, as the gravitation of the whole world" (129). Beliefs exert pressure, just as pressure is exerted on a baby's head during delivery, on Tristram's nose, or on Toby's old wig: "Squeezed up in the corner of his old campaign trunk . . . it was not so pliable a business as one would have wished" (547). The spring of the hair is not as reliable as the spring of the air.[113] Mrs. Wadman presses Uncle Toby's hand "with the gentlest pushings, protrusions and equivocal compressions, that a hand to be removed is capable of receiving," as well as "[making] him sensible, that it was her leg (and no one else's) . . . which slightly press'd against the calf of his" (505). Pressure is also exerted in the opposite direction as resistance to movement. Sometimes things leak as a result—Susannah's memory, for example. Friction describes the workings of the "hobby-horse," as well as numerous stories of rubbing and its effects. Most notable as a force, though, is velocity, a condition that memorably characterizes "some [roasted] chestnut, of more life and rotundity than the rest . . . put in motion . . . sent rolling off the table" and into the breeches of "Phutatorius." Velocity is the condition of Tristram himself throughout volume 7: "The post boy gave a crack with his whip—off I went like a cannon, and in half a dozen bounds got into Dover" (433).

What holds these sequences together is the way that they trace the immediate contact of one body with another somewhere between the initial impulse and a final goal or resting place. Intermediate causes are carefully detailed, and the largest possible contexts or systems of explanation are called on to justify them, but any more comprehensive answer to the question *why* is harder to come by. The narrator jokes that "The Fates, who certainly all foreknew of these amours of widow Wadman and my uncle Toby, had, from the first creation of matter and motion . . . established such a chain of causes and effects hanging so fast to one another, that it was scarce possible for my uncle Toby to have dwelt in any other house in the world, or to have occupied any other garden . . . but the very house and garden which join'd and laid parallel to Mrs. Wadman's" (501). Physical contiguity and causality become scarcely distinguishable. The description of movement appears to be given for its own sake, or to elaborate the laws of motion: on his way to get the man-midwife, Obadiah on horseback moves "at a monstrous rate . . . a phaenomenon, with such a vortex of mud and water moving along with it, round its axis," when he suddenly encounters Dr. Slop at an angle of the garden wall, only "threescore yards from the stable yard." Because Dr. Slop instinctively crosses himself, he "let go his whip"; as a result, he "lost his seat" and fell off his horse: "The Momentum of the coach-horse was so great, that Obadiah . . . rode in a circle three times round Dr. Slop," before he could stop, bring him into the house and rap on the door (94). Walter Shandy cannot stop mentally calculating the time between the initial dispatch of Obadiah on this errand, and the rap on the door, an interval that the narrator informs us is two minutes and thirteen seconds, if measured scientifically with a pendulum (92). This account of Obadiah's movements is given by Tristram to defend himself from an imaginary critic's charge of improbability, of having written "a profess'd Romance" (93). The almost unhinged precision with which Obadiah's trajectory is described, and the sequence of events that then ensues, identifies this novel, then, as a realist novel. It shows how motion occurs mechanically at the level of matter, that is, from "ordinary, comprehensible mechanical and material causes," but it leaves larger questions unanswered.[114]

In Sterne's novel, mechanical devices and movements provide common metaphors. Tristram argues, for instance, "An eye is for all the world exactly like a cannon." Yet he cannot help pursuing this metaphor by referring to an actual cannon and then making the actual cannon into an illustration of abstract laws: "It is the carriage of the eye—and the carriage of the cannon—by which both the one and the other are enabled to do so much execution" (524). The locution of the "plan" seems to capture the material, metaphorical, and abstract qualities involved in every sequence. Plan also names both map and plot. This represented world seems to be largely ruled by friction, the retardation of motion, the kind of "external causes" that we are supposed to ignore. Yet the novel persists in referring to abstract quantities of matter and force, as in the narrator's weary recognition that "upon this page and the five following a good quantity of matter [must] be inserted, to keep up that just balance betwixt wisdom and folly" (559). The phenomena of force and motion, whether the swinging of an arm/pendulum or the "equivocal compression" of Mrs. Wadman's body, are real, localized, and consequential up to a point. But they are also abstract vectors and quantities that make reference to an external world. Significantly, when Uncle Toby is asked the all-important question, "And whereabouts . . . did you receive this sad blow?" (580), instead of pointing to the location on his own body, he points to the map of the Namur: "He could at any time stick a pin upon the identical spot of ground where he was standing when the stone struck him: this struck instantly upon my uncle Toby's sensorium—and with it, struck his large map of the town and citadel of Namur" (580). The motion of the stone strikes Toby's body, his thoughts, and, in a small but astonishing metaleptic moment, seems to strike the map itself. The external world is far more than an inert backdrop to action. It provides the laws that are then twisted into plot and pinned down to the ground.

CHAPTER 3

Defoe's Outstretched World

How plotless can a novel be and still be considered a novel? The case of Defoe suggests that plotlessness was not an accidental flaw in some works but critical to the novel's institutionalization after the start of the nineteenth century. Walter Scott produced the first collected edition, *The Novels of Daniel Defoe*, in 1809–10.[1] Homer Obed Brown notes that "Defoe was praised *or* criticized for the 'plotlessness' of his novels, which are frequently described as a succession of adventures lacking any fixed direction." But in Scott's view, that plotlessness seems to have been a strength rather than a deficiency. "For Scott, what makes Defoe realistic is his lack of coherent 'story,'" Brown argues. Scott saw this very lack of plot as the link between Defoe and Austen, authors who are still often considered the endpoints of the novel's supposed rise. "No two novelists probably seem less alike than Defoe and Austen to modern readers, yet Scott was not the only reader to associate them with each other," in Brown's words.[2] Ruth Yeazell reminds us that Scott compared the novels of both Defoe and Austen to "Dutch painting."[3] Hippolyte Taine returned the favor on their behalf, comparing Scott's novels disdainfully to Flemish paintings: "Ask a cook which picture she prefers in the museum, and she will show you a kitchen, in which the saucepans are so well done that one is tempted to dip into the soup."[4] Novel readers, it seems, were just as happy to dip into the soup rather than following a story as it unfolds.

Walter Scott and Anna Laetitia Barbauld did much to institutionalize the novel through their writings and edited collections: the *Ballantyne's Novelist's Library* (1821–24, 10 vols.) and *British Novelists* (1810, 50 vols.) respectively. But in her prefatory essay to the first volume, "On the Origin and Progress of Novel-Writing," Barbauld complains about how some novels tend to be *too* well-plotted, using Fielding's *Tom Jones* as an example: "Every such work is a *whole*, in which the fortunes of the personages are brought to a conclusion . . .

every incident . . . is introduced for a certain purpose, and made to forward a certain plan. . . . But real life is a kind of chance-medley . . . in real life our reasonable expectations are often disappointed; many incidents occur which are like 'passages that lead to nothing' . . . the reader of a novel . . . would often guess wrong if he were considering the real course of nature."[5] By 1810, it seems that novels can either be well-plotted, with "a structured rising sequence and clear resolution," or they can be realistic.[6] Few would complain that Defoe plotted his novels too well.

But Defoe knew how to frame a plot. In his 1704 account of a destructive week-long storm that blew across England, Defoe includes dozens of memorable micronarratives like this one, efficiently edited and artfully turned:

> One Mr Simpson, a Scrivener being in Bed and fast a-sleep, heard nothing of the Storm; but his Family being more sensible of Danger, some of them went up, and wak'd him; and telling them their own Apprehensions, press'd him to rise, but he too fatally sleepy, and consequently unconcern'd at the Danger, told them, he did not apprehend any Thing; and so, notwithstanding all their Persuasions, could not be prevailed with to rise; they had not been gone many Minutes out of his Chamber, before the Chimneys fell in, broke through the Roof over him, and kill'd him in his Bed.[7]

In five miniature scenes and a handful of telling details, Defoe builds a complete tragic plot.[8] There is a protagonist who makes a mistake and suffers a reversal, "a change from one state of affairs to its exact opposite . . . in conformance with probability or necessity."[9] Each word inches the plot forward until it ends with a dramatic (literal) collapse.

Why, then, do Defoe's novels follow such a different method of story-construction? If the story of the unfortunate scrivener focuses only on the most critical details and events, the plots of his best-known novels, *Robinson Crusoe*, *Moll Flanders*, and *Roxana*, are strikingly repetitive and loosely organized. It is harder than it should be to single out what Auerbach calls "the great exterior turning points and blows of fate" because there appear to be too many of them.[10] The relative importance of each major event keeps changing. Between moments in which our attention is called to the notable workings of fate or providence, there are long stretches in which too many things, too many decisive events, happen too many times. The very title page of *Moll Flanders* (1722) tells us how many times she was married: "Five times . . . whereof once to her own Brother," and the brother was not even the last one.[11] In *Robinson Crusoe* (1719), Crusoe calls more than one of his multiple voyages "the unhappiest voyage that ever man made."[12] Every storm, in fact, seems to be the worst one that ever happened. The first storm happens on Crusoe's very first time aboard a boat, and even on that brief trip from Hull to London, there are not one but two separate, terrible storms, the second of which makes the ship sink to the bottom of the ocean. It would be as though the unlucky Mr. Simpson rebuilt his house every week only to have a storm destroy it every time. Even the unique event, such as the discovery of the footprint, has multiple iterations in multiple modalities. Canoes full of footprint-leaving strangers "might have been there once, or twice" in fifteen months, or many times over the years, as the heaps of bones on the other side of the island testify (146, 130). In their respective novels, Moll Flanders's and Roxana's serial relationships, often with the same man, are hard to keep track of (and the lack of proper names does not help, either). Actions that did *not* occur are also represented as highly consequential events, as when Crusoe ends up not realizing his plan of "killing twenty or thirty naked savages" (135).

In Defoe's novels, the decisions of protagonists anticipate, reflect, and intricately coordinate with a world of larger forces. That world never stops moving. In this chapter, I want to look at how Defoe's plots construct their objectivity on the assumptions and practices of experimental philosophy discussed in the previous chapter. His plots build on this world of intermediate causes, and operate on various scales. At the usual level, plot consists in praxis, or the actions that characters undertake. But there's another more fundamental level at which things are always supposed to be happening—in the background, or between scenes. We might think of this component as the physical, natural, or phenomenal plot. Defoe's novels conceive events in terms of forces that push, pull, move around, or stabilize the pieces of the world: wind, for example, is the protagonist of *The Storm*. This physical plot underlies the actions, emotions, and reflections of the characters. Sometimes it reinforces them, but more often undercuts them repeatedly, taking away something that was thought to be settled. Even social and economic situations such as Moll Flanders's struggle for security or Roxana's struggle for freedom from a past identity feel like they proceed from an aspect of their worlds deeper, more fundamental, relentless, and impersonal than even economic necessity or social institutions such as gender, nation, or class. But it's too abstract, perhaps too subjective or attitudinal, to call this aspect fate or destiny. It seems to reside in the intricate physical nature of the world itself in these novels. Things keep moving, once set in motion. Actions are often understood only retroactively, if at all. Actions rely for their very definition on the way that consequences play themselves out against and within a world. This world includes chance and yet follows certain laws. Causality is both upheld and questioned, tested and respected. This seems to me to lie close to the core of Defoe's realism.[13] Defoe's descriptions often illuminate the consequences of actions undertaken blindly: for instance, sticking a stake into the ground, or shaking out a bag to empty it. The consequences—a tree, fence, crop—unfold according to their own supposedly objective rules.

Defoe's plots highlight the mechanics of movement and the reciprocal nature of physical action in the world. His plots owe their bulkiness and their dramatic interest to the concept of intermediating physical causes: in s'Gravesande's words, phenomena whose "action depends upon another motion," and which in turn produce other observable movements.[14] The plot often focuses on the way something happened, the steps or means by which something occurred. Even in *The Storm*, stories of "remarkable Deliverances" emphasize the ingenious ways through which certain people were spared. In one example, "In the Fall of the House two great Spars seem'd to fall so as to pitch themselves on an End, and by that Means to support that other Part of the House which adjoined to the Upper Chamber."[15] It is *how* they were saved, the specific configuration of the means, that Defoe stresses. Secondary or "immediate" causes seem to hold the key to a good story; they produce narrative challenges and opportunities. Carving out such a realm of action by no means eliminates for Defoe other domains of causality. It doesn't rule out the operation of providence or fate. Those other domains remain available and can be drawn on in order to raise the emotional or affective temperature of the narrative. But the idea of a physical world of matter that moves according to certain laws becomes the basis of a more detailed, temporally extended, densely woven fictionality—what could be called an "ontological realism."[16] The properties of matter—weight, position, resistance, inertial movement—provide the form the action takes, and shape the consequences that ensue. They provide an alibi for action in the sense of an elsewhere, an otherness, as well as a justification for it or even an implied innocence. Particularly for Crusoe but for Defoe's heroines as well, action consists at its base in the application of force to things, either to move them, or to keep them from moving. The plot integrates its actions as tightly as possible into the setting or physical situation; indeed, that physical configuration shapes, justifies, elicits the measure taken. We can think of the famous footprint in the sand that Crusoe discovers: what allows that event to happen, what gives it a shape and form, is the meeting of a foot with sand at just the moment when the latter had a particular consistency. The ratio of water to sand, the holding back of the wave that might have washed it away before Crusoe came by: these circumstances cannot be separated from the event itself, or from his response to it.

Defoe also makes sure that we notice the inherent temporal uncertainty of action when presented as part of a novelistic plot. Actions are deeply recursive, based on concrete anticipation of the future and a constant looping back in time to reconsider what has been done. Any action thus negates both a past and a future that had been considered only in abstract or general terms. This puts time into plot at a different, micrological mimetic level. Over time, sequences of actions stretch or dissolve, are named or renamed. But this comes about because of the attention given to the circumstances and the consequences of even the simplest physical gesture or decisions. Thus, I do not see Defoe's minute, extended accounts of actions as narrative filler whose primary importance is formal or rhythmic, as Franco Moretti has argued.[17] Rather, this feature of Defoe's plots points to something important about how action is conceived in the novel. It's defined by its circumstances and its consequences. In other words, an action is what brings about consequences in the given world but only in elaborate conjunction with material objects, media, and other mechanical forces and causes that are part of the represented world. It is hard to say when an action occurs or is ever perfected, and I see this as a deliberate feature of Defoe's plotting. Defoe's plots constantly refer back to themselves, to things that had happened or been anticipated earlier.[18] This is a critical feature. By referring to actions that happened before, or focusing on their concrete effects, the plot circumscribes actions and yet dissolves their boundaries in time and space. Their meaning and their effects seem to reside in different dimensions. We can see the beginnings of what Barthes called "l'effet de réel," but it takes up residence here in the domain of plot and in the logic of action rather than of description.[19]

The popularity of Defoe's *The Life and Strange Surprising Adventures of Robinson Crusoe, Mariner* (1719), which continues to traverse languages and continents into the present age, must be due in large part to its plot.[20] This is a plot that many people know, even those who haven't read the novel: shipwreck on a remote island, the building up of a life through diligent labor and ingenuity, external threats warded off, and eventual rescue. When reduced to this abstract scheme, Crusoe's plot becomes endlessly transferable to other times and places; it also becomes allegorizable. Captured in a single glance as a coherent shape, it invites interpretation. Thus, Crusoe has been studied as homo economicus, as spiritual pilgrim, as colonialist, as embodiment of the mariner's craft; and the novel has been analyzed for its religious, historical, and sociocultural significance.[21] But can a desert island count as a plot? I think it does: what we might think of as the setting defines the contours and significance of the actions taken within it. Indeed, it places action under unusual scrutiny, and subordinates it in almost every way to the physical world. Nearly every action is described in step-by-step detail. This way of proceeding introduces a gap between what things mean, and how they happen. Interpretation is about what things mean, or *why* rather than *how* they happen. Why did the ship not sink to the bottom? This question differs importantly from the question of *how* it was possible for it not to sink, and for Crusoe to get up into it.

The former dimension, what things mean, is addressed through what Defoe likes to call serious reflection. It is taken up at intervals as a type of spiritual exercise. Regularly, Crusoe reflects on what certain events or situations mean—for example, the fact that he alone survived the shipwreck. But the latter dimension, how things happen, I would argue is the plot's unremitting concern, and we might say that it encompasses the former in the way that the ocean surrounds his island. Crusoe's most heartfelt sentiments are elicited when he interprets the granular *how*, the particular way in which something happened or didn't happen: how the ship remained accessible after its wreck, for example, or how he happened to land on the side of the island where other canoes didn't come. This way of noticing how things happen is not limited to the actions that Crusoe performs. Crusoe often makes reference to something that doesn't yet exist, or a situation that could come about. Even his famous apostrophe to money, for example, queries how it could cause other things to happen: "O drug! said I aloud, what art thou good for?" After declaring it useless, he notes that "upon second thoughts, I took it away . . . wrapping all this in a piece of canvass" (47). You never know when it might become useful; it usually does. The domain of narratability is defined not only by causation in the past, but by the possibility of making something happen through material means.[22] Defoe's novels seem mostly untroubled by a need to distinguish between actions and events, between conscious persons and unconscious things, or to assume that one class should naturally have priority over the other.[23] The reason, I will argue, stems from what Defoe shares with experimental philosophy: a keen interest in intermediating causes and physical forces that appear to operate independently of the observer.

Defoe was well-versed in early modern science. His teacher, Charles Morton, incorporated into the program at his Academy for Dissenters the new sciences to which he had been exposed during his own time at Oxford.[24] Defoe valued his instruction there enough to recommend English (as opposed to Latin) instruction in experimental science as the basis of an ideal education for young men of the middle and upper classes.[25] Defoe's own familiarity with this body of knowledge and practices is demonstrated in his *General History of Discoveries and Improvements in Useful Arts* (1725–26). This history culminates in "the discovery of the magnet or loadstone," which Defoe credits for the present (and future) European domination of the world through navigation, trade, and colonization. Though he could have simply given a summary of the history of the compass, Defoe goes into great detail, citing numerous experiments with magnetism made by "the Learned Mr Boyl [*sic*] and others."[26] He uses the same format as Desaguliers, Goldsmith, and others: for example, "V. If a Magnet be cut through the Axis, the parts or Segments of the Stone, which before were joined, will now avoid and fly each other," or "XIX. Mr. Boyle found that by heating a Magnet red hot, it would be speedily deprived of its attractive Quality."[27] Most of the experiments describe the way that matter will behave when manipulated in certain ways: how magnetic phenomena and movements occur or vary under certain conditions: "XXII. He observed that well tempered and hardened Iron Tools, when heated by Attrition, turning, fileing [*sic*], etc. would while warm attract thin Fileings or Chips of Iron and Steel, but not when cold."[28] The final numbered experiment applies Newton's third law, while succinctly distinguishing magnetism from gravity:

> XXX. Because it is one of the universal Laws of Nature; that Action and Re-action are always equal; therefore it is plain, the Iron must attract the Magnet as much as that doth the Iron, and so you may easily experiment it to be in Fact if you place a Magnet or piece of Iron on a piece of Cork, so as that it may swim freely in the Water, for then you will see that which soever you hold in your Hand will draw the other towards it. From all which Experiments 'tis plain (as Mr. Boyle concludes) that Magnetism doth much depend upon mechanical Principles, as also that there is such a thing as the Magnetism of the Earth, . . . but Sir Isaac Newton demonstrates that Gravity is a very

> different thing from Magnetism, since the former is always as the quantity of matter attracted, but Magnetism by no means so.[29]

Defoe stresses that both forces operate in calculable, predictable ways at the level of "mechanical" causes and effects. He is not troubled here by deeper causes, or the inability of human knowledge to penetrate these. It is enough that the movement of the compass is reliable and allows navigators to move purposefully across the globe. Defoe cannot praise this invention enough, nor does he tire of tracing its consequences.

We recall that how things happen is what Boyle had designated as the proper object of scientific curiosity: "Not . . . what is the general agent that produces a phenomenon," but "*By what means, and after what manner, it is produced.*"[30] Not the question of who made it happen, in other words, but of how the effect was produced. If we don't know how some agent might actually "operate to produce effects," that agent doesn't belong to the world of "natural phaenomena."[31] Of course, one of the first things to be noticed about Defoe is how often he invokes the agency of special providence. There's always the possibility that someone's choices, decisions, or even physical motions were prompted by a transcendent power. But even in such cases, providence operates through mechanical, material, perceptible means that are often described by the fiction in some detail. H. F., the narrator of the *Journal of the Plague Year*, sees the words on the page of the Bible that he opens and consults for guidance. Moll Flanders thinks she hears the devil's voice audibly prompting her to commit her first theft. Crusoe notes that "whenever I found those secret hints, *or pressings of my mind*, to doing, or not doing any thing that presented; or to going this way, or that way, I never fail'd to obey the secret dictate; though I knew no other reason for it, than that such a pressure, or such a hint hung upon my mind" (139, emphasis added). Is pressure here a metaphor, or a literal description of what he feels? It seems designed to bring even the mind or will as much as possible within the range of natural phenomena.[32] Defoe can suggest that the decision of the mind, the impulse to walk in one direction rather than another, may arise from something else that presses on it. There may be gaps in the known chain of material causation: how a hint from providence creates a felt pressure in the brain, or how a decision taken by the will makes the muscles move. Defoe tackles such gaps through a certain strategy: the recursive narration of action, going back in time to subdivide actions into ever-smaller sequences, as I will discuss below. But what allows him to proceed in this way is a tenacious focus on objective causation as the proper domain of both narrative and science.

In his early work, *The Storm*, Defoe seems particularly fascinated by the obscurity of its causes, on the one hand, and the dramatic obviousness of its consequences, on the other. He begins with a long discussion of the challenges that science faces in trying to explain the cause of wind. Defoe ponders the relation between scientific investigation and the attitude of spiritual or religious wonder.[33] This relation is not based solely on different classes of objects, but rather on the degree of diligence, energy, ambition, and success with which someone might pursue their investigations. The philosopher's goal is "Natural Knowledge, *meerly as such*; for 'tis a Natural Cause they seek, from a General Maxim, that all Nature has its Cause within it self."[34] He speculates that King David, the writer of the Psalms, "Was upon the Search of Natural Causes, and found himself puzzled as to the Enquiry, when he finishes the Enquiry with two pious Ejaculations . . . David may very rationally be suppos'd to be searching the Causes, Motions, and Influences of Heavenly bodies; and finding his Philosophy a'ground . . . he turns it all to a pious Use [and] recognizes Infinite Power."[35] Such sublime emotions tend to come into play, he suggests, when science has failed to complete its own task. It is "not enough for [the philosopher] to know that God has made the Heavens, the Moon, and the Stars, but must inform himself where he has plac'd them, and why there. . . . Tis not enough for an Anatomist to know that he is fearfully and wonderfully made . . . but he must . . . search into the Method Nature proceeds upon in the performing the Office appointed, must search *the Steps she takes, the Tools she works by* . . . some things are so plac'd in Nature by a Chain of Causes and Effects, that upon a diligent Search we may find out what we look for."[36] It's true that some natural phenomena are easier to explain than others: "In Some of the Principal Parts of Nature . . . Things appear both in their Causes and Consequences, Demonstration gives its Assistance, and finishes our further Enquiries." Others are harder to dissect. It is "where we see Effects but cannot reach their Causes" that philosophers may find themselves tempted simply to invoke "the Mighty Hand of Infinite Power, the Author of Nature, and Original of all Causes."[37]

But Defoe insists that to do so is to fall short as a natural philosopher: "We may at any time resolve all things into Infinite Power, and we do allow that the Finger of Infinite [*sic*] is the First Mighty Cause of Nature her self: but the *Treasury of Immediate Cause* is generally committed to Nature; and if at any Time we are driven to look beyond her, 'tis because we are out of the way: 'tis not because it is not in her, but because we cannot find it."[38] Infinite power on the one hand; immediate or intermediate cause on the other.[39] While Desaguliers had simply asserted that natural philosophers "do not reason about these hidden Causes," Defoe goes somewhat further.[40] He delineates a realm of intermediate causation that is the proper focus of human investigation, activity, and explanation. It is also the focus of the novel's storytelling. This is what we should be looking for: the steps, the tools, the chain of causes and effects contained within nature.[41] These are plot's own materials.

Defoe is more often struck by the neat contrivances or timely interpositions of intermediate causes than by the abstract concept of infinite power. Such causes seem to fill that troublesome gap between an invisible world and a material one, a vague mental or interior world of wishes and intentions and an objective possibility. Toward the beginning of the second part of *Robinson Crusoe*, the narrator is secretly longing to go back to see his island when his nephew happens to propose just such a journey to him: "Nothing can be a greater Demonstration . . . of the existence of an invisible World, than *the Concurrence of second Causes, with the Ideas of Things, which we form in our Minds*, perfectly reserv'd, and not communicated to any in the World. My Nephew knew nothing how far my Distemper of Wandering was return'd upon me, and I knew nothing of what he had in his Thoughts to say."[42] The best-known example is the bag of seed-corn in the first part of *Robinson Crusoe*. When he notices the English barley growing on his island, he believes the stalks to be "the pure production of Providence for my support": "I began to suggest, that God had miraculously caus'd this grain to grow without any help of seed sown" (63). Soon, though, he recalls that "I had shook a bag of chickens meat out in that place," and comes to the conclusion that "all this was nothing but what was common" (64). But Crusoe then reminds himself that Providence had still acted by means of the "10 or 12 grains of corn" and his act of emptying the bag in the right kind of place for its germination. Boyle had used the example of grinding grain; Defoe's example shows that if an angel wanted to grow grain, he would still need to use seeds and shade to do so.

The "treasury of immediate cause," then, provides Defoe with unlimited amounts of narrative material from the physical world. His *General History of Discoveries and Improvements* opens with an echo of Boyle's description of mechanical philosophy, along with Bacon's emphasis on practical application: "How, and by what Steps these Discoveries were made . . . when, and by whom; to what Uses the Discoveries that have been made have been apply'd to . . . how much farther such Improvements may probably be carry'd: These will be the Subject of our future Enquiry in these Sheets."[43] But it's one thing to explain the steps by which discoveries were made, and another to turn them into a plot. How, then, does Defoe construct plot out of this domain of intermediate causes? What place is there for human action? Defoe's novels tend to construct action retrospectively or even retroactively: by narrating the consequences first and then deducing where and when the action occurred. For example, in the episode of the seeds, the sequence in the narrative discourse goes as follows: Crusoe tells us that while "rumaging [*sic*] my things, I found a little bag" apparently containing only "husks and dust," and empties it out to reuse the bag. After the rains, he sees something sprouting, forgets that he had "thrown any thing there," discovers the English barley, and concludes that it's a miracle ("I knew not how it came there"), until he recalls later that he had emptied the bag in that spot. The pieces are all laid out—the bag, the seeds, the shade, rain, and time, but Crusoe realizes that he had performed the act of planting only *after* he identifies what he sees (green stalks) as consequences, and traces them laboriously back to their cause. His action only emerges and becomes nameable as a result of the whole chain of material consequences that followed upon it. Those consequences, in turn, required particular circumstances that could only be provided by the environment. This could stand as a model for the way that realist plot relies on the concept or figure of intermediating causality as a way to link together elements over time.

Plot in the novel depends on the complex relation that action bears to the tracing of chains of causes and consequences. In *The Storm*, Defoe writes that "'tis a Natural Cause [Philosophers] seek, from a General Maxim, that all Nature has its Cause within it self: 'tis true, 'tis the Darkest Part of the Search, to trace the Chain backward; to begin at the Consequence, and from thence hunt Counter . . . to find out the Cause: 'twould be much easier if we could begin at the Cause, and trace it to all its Consequences."[44] We might say that plot is the very darkness that allows such a thing to happen. As such chains are traced both forward and backward, from a perceived effect or event to its cause, from cause to consequence, actions begin to emerge. Sometimes the name of the action sequence is identifiable only in retrospect, as in the case of Crusoe's planting grain.[45] At other times, and more frequently in the other novels, actions fail to produce their intended consequences: Crusoe builds a canoe, but can't move it because it's too heavy.[46] He brings back grapes to store and use, but finds them all crushed.[47] His failures are numerous and significant because they locate action somewhere along this rather shaky chain of material causes and consequences. Does it count as an action if the clay pot ended up collapsing rather than holding its shape? Action and consequence depend on each other dialectically; we learn from the novel to anticipate both forward and backward, to see not only what will happen, but what might have been the cause for something we now see.

This model of action differs somewhat from the Aristotelian one. Aristotle's model of tragic plot or *muthos* relies on a dialectical sense of a whole that both is and is not realized. The action or plot as a formal structure must exhibit harmony and wholeness through its impeccable organization. In Amélie Rorty's words, muthos "reveals the inner logic and causal organization of an apparently disconnected series of events."[48] But this formal or causal wholeness is negated by what happens with regard to the human intentions and actions that are foregrounded by dramatic representation. "Tragedies represent the way that the protagonist's serious actions—those that affect the major directions of his life and that determine his happiness—skew the essential ends of what he does, and how this error, this waywardness brings disaster . . . they show how the protagonist's (well-intentioned but mistaken) purposes miss the true or essential ends of his actions and how his *hamartia* brings disaster."[49] The model for human action in tragedy is *proairesis*, or decision of a particular type, a commitment that unleashes an apparently inevitable series of consequences.[50] Defoe is clearly aware of this traditional dramatic model of plot. Characters such as Moll Flanders or Roxana repeatedly face situations in which they must make a choice or decision. In Jean-Pierre Vernant's account, "*Proairesis* . . . rests upon . . . a rational desire . . . informed by intelligence and directed, not toward pleasure, but toward a practical objective that thought has already presented to the soul as a good. Proairesis implies a previous process of deliberation. . . . At the end of this process of reasoned calculation it sets up a choice . . . that is expressed in a decision that leads directly to the action."[51] Though with a great deal more looseness, delay, hand-wringing, and repetition, this is more or less what Defoe's heroes and heroines do.

The important change that Defoe makes in the nature of fictional plot isn't concerned with the inner structure of decision, which he retains. In *Robinson Crusoe*, rather, it has to do with the relation between action, knowledge, and time, the scale at which they are shown to interact, and, most of all, the specification of intermediating causes as things that produce reliable, precisely calculable physical effects in the external world. This is highlighted by the single classic recognition scene that occurs in the first part of *Robinson*

Crusoe. In this novel of adventure, distant travel, and long separations, full of potential for dramatic recognition, the only such scene is given to Friday. After several years living with Crusoe, Friday discovers his father held captive in a canoe left behind after a battle. Friday's extravagant gestures are described at great length: "Friday kiss'd him, embrac'd him, hugg'd him, cry'd, laugh'd, hollow'd, jump'd about, danc'd, sung, then cry'd again, wrung his hands, beat his own face and head, and then sung, and jump'd about again" (187). One action follows another in this account, but even the idea of action takes on a theatrical quality here because of the lack of intermediating causes or material consequences. Notably, there is no need for any material token or process of identifying or reasoning: Friday knows his father instantly and without the intervention even of a material sign. Terence Cave has argued that *anagnorisis* in the postclassical context takes on a somewhat disreputable character, and that seems to be the case here.[52] It is true that Moll Flanders, too, undergoes an unwelcome recognition when she discovers she has married her own half brother by accident; and Roxana spends most of the last part of her novel trying to avoid such recognition by her own daughter. Both of those plots follow in part the Aristotelian model: the characters think they're making rational choices that will ensure their future well-being, but end up facing disaster. Yet even in those plots, the waywardness seems to reside less in the characters' actions than in the material conjunctures of a world in which things never stop dwindling or shifting their places. In *Robinson Crusoe*, we see a contrast between this kind of dramatic recognition, apparently suitable for characters like Friday, and Crusoe's plot, which hinges on a different kind of recognition, mediated at every step by the material characteristics of things.[53]

We see this other kind of recognition, for example, in the deliberate experiments that Crusoe makes by sowing his grain at different times of the year: "By this experiment I was made master of my business, and knew exactly when the proper season was to sow; and that I might expect two seed-times, and two harvests every year. While this corn was growing, I made a little discovery which was of use to me afterwards . . . the stakes which I had cut out of some trees . . . were all shot out and grown with long branches. . . . I was surpris'd, and very well pleas'd, to see the young trees grow . . . it is scarce credible how beautiful a figure they grew into in three years" (84). This sequence, technically anagnorisis, illustrates, in Cave's words, "a shift from ignorance to knowledge," but this shift results from conscious manipulation of natural phenomena, the observation of results, and their projection into a future imagined as regular and consistent.[54] Both surprise and aesthetic pleasure are present, but there is nothing on the scale of Friday's joy. Such recognition seems more appropriate for a laboratory than for the stage. Discovery of Crusoe's sort seems to be the opposite of conventional recognition; instead of joining what had been separated or revealing what had been improperly or accidentally concealed, Crusoe's type of discovery follows the model of the scientific investigation of causes and consequences. It prepares the way for further "improvement" or repetition of the same act, both of which Crusoe usually goes on to do. For example, he decides to cut more stakes and plant yet another fence around the perimeter of his original house. Later, after he discovers the footprint in the sand, in an extended state of panic he sets "near twenty thousand" (128) of these stakes all around his outer wall. Such behavior further stretches the temporal boundaries of action: Crusoe's fence-building is never done.[55]

A curious thing about actions and narration in this novel is how intensely recursive they *both* are, at the level of story and of discourse. Crusoe's fall-back activity, when he has nothing else to do, or when some newly registered external presence makes him feel vulnerable, is to fill in his already-impregnable fences and walls. There are many such fences, each with layers upon and between layers. Next to his enclosed "country habitation," for example, he builds another enclosure for his goats. "I had taken inconceivable deal of pains to fence and enclose this ground, so I was so uneasy . . . lest the goats should break thro', that I never left off till with infinite labour I had stuck the outside of the hedge so full of small stakes, and so near to one another, that . . . there was scarce room to put a hand thro' between them, which afterwards when those stakes grew . . . made the enclosure strong as a wall" (121). He had already laboriously fenced and enclosed this ground, but (or therefore) continues to fence and enclose, adding more and more. Around his original dwelling, he adds to his "double row of trees," already thickly grown, more stakes and piles; he thickens the existing wall, then "stuck all the ground without my wall . . . as full with stakes or sticks of the osier like wood . . . as they could well stand" (128). And in yet another enclosure: "I continu'd to perfect the fence till I had made it as secure as the other" (129).

The narration performs a similar operation, going back again and again to the same incident, as every reader will notice. But each time, it adds more detail about the smaller actions that were purportedly embedded within the larger action. In the first iteration of the fence-story, Crusoe "pitch'd two rows of strong stakes, driving them into the ground . . . took the pieces of cable . . .

laid them in rows . . . placing other stakes in the inside" (48). But after this account he gives another one of how he barely managed to cut the stakes in the wood and bring them back, acts that were left out of the first account. Likewise, in his journal, he notes that on Nov. 17, "I began to dig behind my tent into the rock to make room," but only then notes that he didn't have a shovel or spade, which was "absolutely necessary" to perform this task. In order to make the shovel, he has to go find a hard material; to get that sufficiently hard material he has to cut down the tree that will provide it, and so on, in an almost Shandy-esque backward movement. Franco Moretti has designated these detailed accounts as "filler," alluding to Roland Barthes's model of narrative: "A cardinal function is a turning point in the plot; fillers are what happen *between* one turning point and the next."[56] Through their rhythm and small-scale focus, Moretti argues, filler offers a "new, truly *secular* way of imagining the meaning of life: dispersed among countless minute events . . . tenaciously *there*."[57] But Crusoe's accounts appear to suggest that these intermediating events were themselves turning points: without making the shovel, he could not have dug his cave. Without cutting down the trees and lugging them back, he could not have built his fence. It's not about the meaning of such events so much as their causal necessity, their indissoluble connections to each other as cause and consequence. Defoe wants us to believe that every act in this sense (even the nth one) is consequential, a link in a chain. That chain is to be considered forward and backward. Each tree, each stake adds one more necessary piece in the micrologic of plot, the story of how things happen through intermediating causes.

There is a tendency in *Robinson Crusoe* to keep going back and forth like this, to jump forward to specify the possible consequences or future utility of an action, and to jump backward in time either to tell or to complete an earlier action. This is important because it seems to testify to the existence of gaps through the process of creating them (through anticipation) or filling them in retrospectively. And I want to suggest that these gaps exist where the level of causation intersects with narration. Recall Desaguliers's example of intermediating causes: "A heavy Body by its Descent moves the Axis of a Wheel, that carries round another by its Teeth, which by the Intermediation of other Wheels and Pinions, carries round a Hand upon a Dial-Plate to measure Time, or for other Uses."[58] The movements of Defoe's narrative discourse are often devoted to filling in these intermediating causes: what had to happen, or be made, or moved, or put in a certain place for something else to occur. Intermediation can also occur by way of human characters. Helpers and go-betweens are more than usually vital to the plots of *Moll Flanders* and *Roxana*; in the latter, Roxana's maid, Amy, becomes something like a surrogate or extension of the protagonist. Moll relies on the services of various intermediaries, as does Crusoe. But as such intermediating links multiply, there also increases the chance of something going amiss. Amy vows to murder Roxana's troublesome, pursuing daughter, and possibly does so. Or links can fall out and leave problematic gaps, stalling or halting the action.

This recursive, gap-creating and gap-filling tendency lends a sense of objectivity to the fiction through a kind of infrareferentiality. Both the plot and narration turn and return on themselves. The novel refers to the same thing so many times and from so many temporal directions that the phenomenon comes to have the appearance of existing independently of anyone's will. For example, Crusoe discovers on a different side of the island the human remains of many feasts, and decides to avenge the victims by "killing twenty or thirty naked savages" (135). But he doesn't, and the cause for this is not a change of heart. The change of heart comes after the repeated nonappearance of the cannibals day after day, though he goes out every morning with his telescope to look for them, as he mentions repeatedly ("made my tour every morning . . . kept up my daily tour to the hill" [134–35]). "I began to be weary of the fruitless excursion, which I had made so long, and so far, *every morning in vain*, so my opinion of the action itself began to alter" (135, emphasis added). The daily nonappearance of visitors acquires the weight of an objective cause.

Even though he may not yet know what to call his action, or even whether it amounts to such, Crusoe always tells us the exact steps by which he accomplished something, "by what means and after what manner," in Boyle's words. For example, when he goes back to the wrecked ship, at first he cannot see any way to climb up into it. "I swam round her twice, and the second time I spy'd a small piece of a rope . . . with great difficulty I got hold of it, and by the help of that rope, got up into the fore-castle of the ship" (40). He proceeds to systematically dismantle the ship and bring all of its parts—cables, canvas, masts, everything—to land, to incorporate into his walls, fences, and dwelling-structures. The point I want to emphasize is that, at whatever point they become nameable or recognizable, Crusoe's actions aim to rearrange or move or fix matter in place, tie it down. It seems appropriate to refer to *matter* rather than to discrete objects not only because of Crusoe's admirable repurposings, but because of the way in which the plot itself abstracts or homogenizes them. Again, this seems like a thinking with objects, but of another variety.[59] When Crusoe encounters, transports, and uses things, whether living or nonliving, they present themselves to him in terms of weight, length, size, or shape, density or ductility, and number. Other qualities figure as well, such as the ability of the cut stakes to grow into trees (but the trees function as a material screen, measured primarily in terms of density). What Crusoe does with them reinforces the idea that "all matter is homogeneous, or of the same Nature in all Bodies, whether solid or fluid, hard or soft, more or less heavy . . . the whole Variety of Bodies, and the different Changes that happen in them, entirely depend upon the Situation, Distance, Magnitude, Figure, Structure, and Cohesion of the Parts that compound them."[60]

The physical qualities and objective behaviors of the material world appear to define the temporal and ergonomic parameters of action: how much force is required for how long. On his first trip to the wrecked ship, Crusoe builds a raft; as he does so, he thinks primarily in terms of how much weight it can carry. "My raft was now strong enough to bear any reasonable weight" (41). He loads it, sails it back to shore, and guides it into an inlet, but here it "ran a-ground at one end of it upon a shoal" and its load nearly slides off into the water. "I did my utmost by setting my back against the chests, to keep them in their places . . . holding up the chests with all my might, stood in that manner near half an hour, in which time the rising of the water brought me a little more upon a level, and a little after, the water still rising, my raft floated again" (42). He guides it to a different cove, where he repeats a variant of this operation until he can get it onto a flat piece of ground. (A later raft, too "overloaden" (46), actually tips over, sending all its cargo to the bottom.) This is often how actions are described. Think of Crusoe "holding up the chests with all [his] might . . . near half an hour," waiting for the tide to come in, level, and eventually balance the raft. Action acquires its consequences and even its very contours from the behavior (observed or anticipated) of the material world: Crusoe's back braced against the raft, his legs rooted in the sand. In other words, time in the plot is not simply a matter of order (the order in which events happen, in which they're told), or of abstract duration. Rather, it's intricately bound up with the concrete arrangements and movements of matter and the energy that it takes to move or stop things. Crusoe plants stakes "of that wood which I had set before . . . so that in one year's time I knew I should have a quick or living-hedge" (95). Think of how often plots incorporate scenes of waiting. Crusoe is always specific (and sometimes contradictory) about the temporal scale of action: making two earthen jars by hand is an action that takes two months; firing pots in a crude kiln is a five-to-six-hour action. The importance of this dimension has nothing to do with the social value of the thing produced. Rather, it draws our attention to the natural processes or material behaviors involved, and how they elicit certain kinds of human actions: the stickiness and weight of the clay, the heat of the fire, digging, tempering, kneading, shaping, and so on.

The novel thus acquires a strange, though understated, phenomenological power through its detailed accounts of Crusoe's struggles to move, arrange, or rearrange things.[61] There is, for instance, the first canoe that Crusoe decides to make. He tells us its exact dimensions, and how many days he spent engaged in various steps of the process. But it proves too heavy for him to move, though he tries many methods, including digging a canal to connect the canoe to the water. When he calculates "how deep it was to be dug, how broad, how the stuff [was] to be thrown out," he finds that "it must have been ten or twelve years" to move the quantity of soil required (102). Countless times, Crusoe finds himself helped or hindered by the sheer weight of things, and the resistance that they offer: the roll of lead too heavy to lift, the water that resists the movement of his hatchet. The sides of his clay pots initially fall in or out, the clay "not being stiff enough to bear its own weight" (96). He finds ways to alter the physical structure of these things so that they retain their shapes, stay in place or move only when he wants them to.

It is at this level that we can locate a sense of totality: not at the formal level of plot as shape, but rather through the uniformity of the physical forces that such action encounters or requires in this world. Matter behaves consistently; it needs force to be set in motion or to be stopped. This idea is reinforced in a violent way in the *Farther Adventures*, when Crusoe, traveling across China into Russia, encounters villages that worship a particular "Idol made of Wood." Scandalized, Crusoe describes the people prostrating themselves before this "Scare-crow," "flat on the ground, round this formidable Block of shapeless Wood: I saw no Motion among them any more, than if they had been all Logs of Wood like the Idol" (329).[62] Crusoe becomes violent, even murderous, and wants to "destroy that vile abominable Idol, and let them see that it had no Power to help itself, and consequently could not be an Object of Worship" (331–32). In a midnight raid, he captures and binds its priests and makes them watch while he fills this idol with gunpowder and blows it up; it burns into a log of wood. The lesson is clear, if not exactly subtle.

But even if matter is inert, that doesn't mean that it stays still. It's much more likely to have been set in movement. As an English merchant says, "What should we stand still for? the whole World is in Motion, rouling round and round; all the Creatures of God, heavenly Bodies and earthly are busy and diligent, Why should we be idle?" (248). Crusoe notes in response, "If Trade was not my Element, Rambling was" (248–49) and agrees to go all over Asia and Russia. Not only the character of Crusoe but the physical world itself seems to be moving, even spreading apart. Defoe's *Journal of the Plague Year* (1722) obsessively documents both the spreading of disease and efforts to confine and contain it. Moll, Roxana, and Crusoe expend inordinate amounts of energy trying to find ways to secure their possessions, to keep them from dissipating, getting spent, stolen, or lost. Husbands die, children get displaced or misplaced, things and money get lost unless efforts are made to tie them down. Moll feels herself at the mercy of such entropic forces and constantly expresses the fervent desire to settle, to stop moving in a safe place. All these changes, though, are mostly in situation, distance, and magnitude rather than in some internal essence, which is why making an inventory seems to have an almost performative character. This also explains one of the most reliable impulses shown by Defoe's characters: to draw up a contract for someone to sign. Moll begins and ends her relationship with her husband/brother in this way. Contracts try to bind, inhibit, and ensure the future positions and movements of persons, money, claims, things. The anticipated movements of matter provide one of the more important pretexts or alibis for action.

Defoe's plots emphasize the force required to move things. But the force needed to *keep* things in their places, to curtail further movements and consequences, seems to be more consequential socially and politically. The political-epidemiological drama of the *Journal of the Plague Year* centers on the question of whether shutting up houses was justifiable or efficacious as a civic policy. Because things move unless they're stopped (not only in strict or calculable Newtonian terms but as a general understanding of the social world as well), putting things together and fixing them in their places constitutes not just making but ruling. It is politics in the sense that Rancière terms "police": determining the places where things and people should go—and stay.[63] In the first part of his adventures, Crusoe mostly puts together wood, metal, earth, as well as bonding Friday to himself. When he returns to his island in the second part, his chief accomplishment is to make the European men now living there marry the native women they are living with. Joining these men and women in marriage is presented as one of the most important things he does, and apparently legitimates his rule. He claims that he is without actual "Authority or Power" (217), but they all know that they owe their existence to him. The other main act he performs on this visit is his "Disposition of Things among the People," that is, distributing the goods he has brought for them, and deciding who'll live where, and who'll receive how many of the "Indians" now conscripted as servants. He draws up a contract "setting out the Bounds and Situation of every Man's Plantation, and testifying that I gave them thereby severally a Right to the whole Possession and Inheritance of the respective Plantations or Farms, with their Improvements to them and their Heirs, reserving all the rest of the Island as my own Property, and a certain Rent for every particular Plantation after eleven Years, if I or any one from me . . . came to demand it" (192).

Defoe links novelistic plot to mechanical philosophy's focus on intermediate causes. Plot becomes the story of how things happen in a concrete totality. There's clearly an important role for human action. But actions bear a complex relation to causes and consequences. The latter have to be traced and retraced, and their conditions examined repeatedly. Plot offers an alibi as well as a motive for action. The alibi is that action grapples with the world at a level of molecular resistance, adhesion, and dispersal. It can thus justify itself and create through the plot the sense of a world that exists and behaves independently. Invoking a sense of a world that is obscure but calculable and orderable through effort, a world in which consequences unfold by themselves, Defoe's plots conceive of action in terms of force, resistance, and time. They suggest that rather than offering stories of wars, victories, social rituals and sacrifice, separations, dramatic recognitions and heroic deaths, novels can base their plots on matter, motion, on the force and resistance of the external world. Politics is unquestionably a dimension of every action: not only when Friday places Crusoe's foot on his head or when Crusoe disposes so blithely of land, women, indigenous people, and rights for all time. But power has gone into hiding within the story of intermediate causes, within the idea that action merely responds to the inertial stasis or movement of the material world.

This idea of the physical, external world as a totality links Defoe's novels to the epic as it was theorized at the beginning of the nineteenth century. An action in Defoe's novels "can be brought before us only within an outspread world and demands the portrayal of this world in its entirety," in Hegel's words.[64] Epic "recounts what is and what happens and therefore has objectivity for its content and form": in terms of the story's content, the protagonist's "deed must prove to be conditioned and brought about . . . by his entanglement in external circumstances."[65] Hegel argues that epic "moves in the element of an inherently necessary total state of affairs."[66] By this latter phrase, Hegel refers to a cultural or spiritual condition, one that Lukács develops in his *Theory of the Novel*. But Defoe and novelists writing in the eighteenth century found an "inherently necessary total state of affairs" in the material world that natural philosophers were investigating in various ways. This epic quality of the novel's external world, though, could only come into view as the later genre began to be institutionalized at the start of the nineteenth century.

CHAPTER 4

Place, Type, and Order

Plot as Natural History

The plots of early realist novels rely on a model of the physical world that gives a certain fundamental status to the phenomena of movement and matter. Far from being an inert or chaotic background, this physical world with its laws and properties conspires to give characters opportunities to act, to succeed, and to fail. It shapes sequences and sculpts both possibilities and outcomes. The window-sash falls decisively; the very hot chestnut rolls into a lap; the heavy canoe resists being pushed or rolled to the water despite all the force Crusoe exerts. The movement, resistance, and growth of things appear to possess something of a lawlike character, even if many contingent circumstances are involved in the actual occurrence. Plot *refers* to that regularity and makes it apparent to the senses.[1] This implicit reliance on objective regularities in the external world seems to characterize the realist novel's fictionality.

But my examples so far have been selective. The novel is usually concerned with other things besides the properties of the physical world. Robinson Crusoe only preoccupies himself with properties of matter such as weight, ductility, and density because there are no other people on his island for many years. All he has are things: physical matter to be arranged and rearranged. For most of the novel, the success of his projects depends only on the physical properties of his surrounding world. But most novels, it is true, inhabit a more populous social world, and focus less on things than on people. Once other people enter the picture, different considerations come into play: observations, predictions, apprehensions, and calculations about their nature, as manifested in their likely behavior. Such considerations can be seen clearly in a novel like Defoe's *Moll Flanders* (1722). When Moll (or Roxana) chooses a man to target in some way, often to rob, marry, or save for later, she has to appraise his qualities in more complex ways. She has to predict his behavior based on the type of man he appears to be. Is he the type to marry first and ask about her finances later? Is he the type to wait patiently for her to come back or to run off with someone else? Knowledge of physical laws will obviously not be sufficient in these cases. There appears to be another kind of objective knowledge involved—one on which such appraisals can rely. Here, too, novels put abstract knowledge about natural phenomena back into concrete situations and particulars.[2] But the knowledge in question has to concern itself with the differences and similarities between types of beings rather than the uniformity that can be found in various kinds of physical events. This chapter turns to natural history, another branch of the field of natural philosophy. While natural history has its own history dating to antiquity, I will focus here on the way it was systematized in the eighteenth century.[3] In some canonical novels of the mid-eighteenth century, plot appears bound up with this project. Plot even appears to be a medium in which such knowledge is enacted.

Plot will be considered in this chapter not simply as enchained sequences of actions or events but as a medium of visibility.[4] To put it another way, sequences of actions (the story, in narratological terms) also represent a world in which they occur. In this chapter, plot will be regarded not in terms of a deep structure beneath the events, but as a plane on which actions and events become visible. Plot represents, makes visible, and unifies disparate phenomena. But it doesn't do these things in the same way at all times or for all genres. How plot in the novel works as a medium of visibility will depend on how representation and knowledge are conceived. We can say that generally, in addition to cause and consequence, plot invites us to consider relations of similarity and difference among the phenomena it discloses. Among other things, plot can link, compare, differentiate, prove, and test. Two characters, for example, could ultimately be shown to be alike in some ways but different in others. Or a trial could be successful in one instance and unsuccessful in another. Or there may be a crucial analogical relation between two episodes.[5] But in the eighteenth century, such practices of comparison and the conclusions that followed them had legitimacy as a type of knowledge. As I will discuss in more detail below, natural history in this period relied on bringing together disparate phenomena so that gradations of similarity and difference became clearly visible. To see varieties of animals or plants, for example, arranged in an ordered system was to know them. Such a belief assumes that the external world not only moves but exists in a perceptibly orderly fashion: as species or types. This belief collides with the realist novel in productive ways, as I hope to show.

This chapter will explore what happens when the classical idea of plot as a formal unity encounters this model of knowledge. What results is a narrative deeply committed to the idea of order, and to the realm of the visible. In both Fielding and Richardson, though obviously not in identical ways, plot relies on both of those models: the aesthetic model of plot and natural philosophy's model of a regular, orderly world in which things not only move in regular ways but exist as types in nature. Invoking both of these claims, plot becomes quite powerful as a means of placing phenomena, of locating them within a single, articulated, and potentially meaning-filled order. Plot thus offers itself to the novel as a way of *relating* and *ranking* empirical phenomena, placing them in relation to each other and often in relation to an implicit ideal order. This function is enabled in part by the plot's very fictionality, conceived in Aristotelian terms as distinct from historicity.[6] A novel's plot does not record what happened to one particular being, but has the power to describe what might have happened, could or should have happened. But the novels to be examined here do not go straight to the universal, or to things that exist everywhere and in all times. Very much of their time in this regard, they place, order, and describe phenomena that reside at a kind of intermediate level: as natural types or, as Fielding likes to call them, "species." Individuals exist in relation to the species, and one species in relation to others. This is the case even for a singular individual or surprising event. Likewise, in these novels, we find that the same or similar events occur many more times than we might expect, and each instance is slightly different, reveals a new angle or relation.

If we can make this claim about the novel as a genre—that it relates and ranks empirical phenomena considered as types—it does so not through description alone but also through the operation of plot. In the simplest scenario, through dramatic encounters and experiences (fights and struggles, for example, but also rescues, acts of assistance rendered, investigations, etc.), characters both reveal themselves as implicit types and rank themselves according to certain criteria: physical or moral strength or virtue, for example. The relation of such events to each other within the larger plot can suggest where they belong in an ideal order. This is not a particularly new observation: William Hazlitt, for example, remarks that Fielding "makes use of incident and situation only to bring out character."[7] However, I'd like to examine in more detail what makes this possible. What larger structures of knowledge give particular weight to plot's serial revelation and ranking of character? Hazlitt also suggests a link to natural philosophy: "The extreme subtlety of observation on the springs of human conduct in ordinary characters, is only equalled by the ingenuity of contrivance in bringing those springs into play, in such a manner as to lay open their smallest irregularity. The detection is always complete, and *made with the certainty and skill of a philosophical experiment.* . . . The truth of the imitation is indeed so great, that it has been argued that Fielding must have had his materials ready-made to his hands, and was merely a transcriber of local manners and individual habits."[8] Hazlitt wants to defend Fielding against the charge that the latter merely went around picking up local specimens like a collector of rocks or plants. But he identifies a key context in which Fielding's use of plot becomes possible: the context of "philosophical experiment," as we have seen, and that of natural history. These contexts also figure in the novels of Richardson.

To summarize, plot does more than orchestrate serial causal chains. It offers a unitary medium in which varieties of nature can be exhibited and related to each other in a kind of dynamic array of resemblances and differences. To make this argument, I will consider some of the discourse and practice of natural history, which flourished alongside the novel in the mid- to late eighteenth century. My aim is to show that the novel, in hands as diverse as Fielding's and Richardson's, engaged in similar ways with this project of naming, knowing, describing, and placing or ranking—through plot—the manifold forms believed to exist in nature.[9] The natural historian Buffon recommends a methodology that also applies to the way in which realist plots go about their business, respecting individual differences but looking for a larger order.

The centrality of natural history to the genre of the novel and to the Baconian project has been much studied.[10] Early modern scholars, amateurs, travelers, colonizers, and others pursued the collection, investigation, and classification of nonliving natural things and living creatures, and were enjoined to follow certain habits of observation and documentation.[11] Thus, in relation to natural history, critics have often stressed the novel's rhetorical stance or claim to truth—as in the kinds of details described and styles adopted. Natural history continued to lie at the core of the activities of a large portion of the Royal Society's membership throughout the eighteenth century. Not mathematical sciences, but natural history and medicine, Harold Cook argues, were "the 'big science' of the early modern period."[12] If we look at a volume of the published *Philosophical Transactions of the Royal Society* from 1742–43, for example, the contents range from medical cases and meteorological phenomena to methods and inventions (for example, "A method of making a gold-colour'd glazing for earthen-ware," "A method of preparing specimens of fish, by drying their skins"). A large number of articles and correspondence, however, is devoted to natural historical subjects—seals, eels, and insects, for example. We also find an "Extract of a letter from J. F. Gronovius, M.D., at Leyden, November 1742, to Peter Collinson, F.R.S., concerning a water insect, which, being cut into several pieces, becomes so many perfect animals"—the polypus.[13] This extract notes that the first account of this creature had come the previous year from "Monsieur Buffon, of the Royal Academy of Sciences at Paris, and F.R.S.," who will be discussed in more detail below. Noting, describing, classifying, and experimenting existed on a continuum of natural philosophical practice and knowledge.

But the epistemology and ontological assumptions that guided natural history in this period are relevant not only to description or narration but also to the way that plots do their work. While the later discipline of biology sought to discover "more fundamental laws" regulating "the basic functions of all organisms," natural history in the eighteenth century sought "order in the diversity of animal and plant form, or in the overall classification of living beings," Paul Farber argues.[14] In this way, it also departed from earlier practices of collecting or what Katie Whitaker calls "the culture of curiosity," in which

> the collection itself was governed by no such systematic classification. Natural and artificial objects were displayed together. The curiosities were crammed into a small space where they filled the walls and ceiling. . . . Widely diverse objects were brought into close proximity in collections so that their variety was emphasized and the contrast between them could be better appreciated. This contrast and variety produced wonder in the viewer. The systematic arrangement of objects into kinds according to some method of classification, which was adopted in eighteenth-century natural historical collections, would have been inappropriate in a seventeenth-century curiosity collection, since an object surrounded by similar but slightly different species would have seemed unsurprising.[15]

Eighteenth-century natural history tried to arrange objects together with "similar but slightly different species," as it looked for ordering principles within nature itself *and* trusted that the practice of ordered representation could make these apparent.[16]

The order in question was assumed to inhere in the visible characteristics of the products of nature, without having to be inferred. That is, through proper arrangement, order could be grasped by the eye. Arrangements were based on the precise discernment and careful tabulation of identities and differences. The three branches of natural history—botany, zoology, mineralogy—shared a common goal: "To describe, name, and classify the diverse riches of nature . . . to be described in terms of their natural *species*."[17] In *The Order of Things*, Foucault argues that in what he calls the classical episteme of the eighteenth century, "a knowledge of empirical individuals can be acquired only from the continuous, ordered, and universal tabulation of all possible differences."[18] The "field of knowledge" was organized by "identities and differences" that were thought of as objective, visible, and representable.[19]

Foucault's account can present some features of novelistic plot in a new light. Conversely, thinking about what the novel tries to do can place natural history in a slightly different perspective. Eighteenth-century natural history, Foucault argues, aimed at "complete enumeration . . . categorical arrangement that will articulate the field of study in its totality."[20] It relied on "a new field of visibility," or a new way of looking at things. Things were scrutinized as "lines, surfaces, forms, reliefs," as natural history examined "the extension of which all natural beings are constituted . . . the form of the elements, the quantity of those elements, the manner in which they are distributed in space in relation to each other, and the relative magnitude of each element."[21] (Think, for instance, of the Linnaean system of classifying plants according to their flowers.) Most importantly, knowledge was thought of as *coextensive with "the general element of representation."*[22] To make something appear in a certain way was to know what it was. Representation was, in a sense, the concept within which knowledge lay, and not the other way around. The format or device that Foucault chooses to stress is the table: "A simultaneous system according to which the representations express their proximity and their distance . . . the network, which, outside chronology, makes patent their kinship and reinstates their relations of order within a permanent area . . . the table of identities and differences."[23] For Foucault, the table is an ideal structure as well as an empirical one, a two-dimensional grid of thinking, not just synchronic but outside chronology altogether. Other scholars note that such a structure can be found in the encyclopedias that were compiled by Ephraim Chambers, Diderot, and others, as well as in the natural history museums, collections, and botanical gardens that proliferated in this period.[24] Place functioned both "as an ordering term, a process in which the world and its products were mapped, named, and classified" and "as sites in which collected, named, and categorised phenomena were represented with a view to benefit different audiences."[25] As Peter Bowler puts it, "Enlightenment naturalists were obsessed with the desire to create a system in which every species could be assigned a unique position in a comprehensible pattern."[26]

A novel's plot is obviously not like a table. It unfolds over time, sequentially. Moreover, to say that plots make things visible is to speak metaphorically. Yet I want to suggest that the idea of "a system in which every species could be assigned a unique position in a comprehensible pattern" describes what some novels' plots do, particularly in the case of Fielding. A plot can be analyzed and assessed as cross-sections, as it were; in any moment of time, the various places and relations of elements can be described. Hogarth's visual narratives draw their power from the fact that plot exists both as a slice of time and a sequence. Within each plate of his 1747 series *Industry and Idleness*, for example, we are invited to study the often precarious relations within the elements represented in the picture. But the relations between the plates obviously matter just as much: most of all, the relations of cause and effect, similarity and difference.[27] Fielding also enjoys the rich possibilities offered by the "meanwhile."[28] One character is at the inn, while the other is on the road; one is upstairs in her room, and another downstairs in the kitchen. Richardson, too: Lovelace is in the garden, waiting, while Clarissa is up in her room, writing. Clarissa is hiding; Lovelace is searching. We are painfully held within the meanwhile. At the same time, the sequence of those scenes makes certain things emerge into full knowledge, offers recognition that would not otherwise be possible. Recognition scenes of the classic variety simply bring into more intense focus a dynamic that occurs throughout the plot in a more modulated way. That gradual emergence into visibility is indissociable from a sense of space. Think of the innumerable minor recognitions that occur in the special chronotope of the road, which is equally temporal (an interval of time) and spatial (a marked route from one point to another).[29] Plots in a realist novel are full of places and positions that are located in the fictional world (the upstairs and downstairs of an inn, or the front rooms and back rooms of a London house), which refer to a real world, and yet which also function as a kind of ranking system, a way of indicating something about the nature of the elements included, and their relation to each other. Think, for example, of the metamorphoses undergone by Joseph Andrews in Fielding's novel, who begins in servant's livery, is stripped of it (literally) to reveal his incongruously white skin, and finally assumes his rightful place as a gentleman's son, in full view, properly clad and labeled. There is more to see here than his social rank. We're meant to see his proper nature revealed, and that can only be done by showing him in a certain position.

The dimension of plot in these novels resembles what Foucault called a "homogeneous space of orderable identities and differences . . . a unitary field of visibility and order."[30] It is more than a matter of a novel's taxonomic ambition. The ability of plot to serve as a unitary field rests, rather, on several other important features. The first is the way that events in the plot involving matter and movement observe lawlike regularities. Plot relies on just that stubborn belief, in Hume's words, "that instances of which we have no experience, must necessarily resemble those, of which we have," or "that like objects, plac'd in like circumstances, will always produce like effects."[31] But though the novel will later locate that regularity or order merely in the habits of the mind, as my later chapters will show, that tendency is not yet very pronounced. The realist novel emerged in a period when the objective regularity of nature, of matter's movements, was being actively tested and described, as well as consistently and productively put to use. The novel was well-positioned to consider such regularity *both* as a feature of the objective physical world, *and* as a habit rooted in the mind.

Plot's formal qualities offered another reason that it could be regarded as a unitary conceptual and representational space for the comparison of identities and differences. The Aristotelian model of plot as a single, coherent action turns up frequently in the criticism of this moment. Aristotle is invoked (via Addison) as "the greatest judge of composition that ever was" in Richardson's "Postscript" to *Clarissa*. Although Richardson is mostly preoccupied here with the idea of catharsis, he uses Aristotelian language to refer to the "catastrophe" of his novel.[32] Tellingly, Richardson tries to excuse his novel's inordinate length with the suggestion that "the letters and conversations, where the story makes the slowest progress, are presumed to be *characteristic*. They give occasion likewise to suggest many interesting *personalities*, in which a good deal of the instruction essential to a work of this nature is conveyed." He hopes "that the characters are various and natural," and concludes that "*if* there be a variety of incidents sufficient to excite attention . . . the length then must add proportionably to the pleasure that every person of taste receives from a well-drawn picture of nature" (1499, original emphases). A well-drawn picture of nature's varieties, with an emphasis on minute gradations, is very close to what natural history aspires to do, as will be seen in the work of Buffon.

Fielding hews more closely and enthusiastically to Aristotle's *Poetics*, emphasizing especially the privilege the latter accords plot as the most important feature of mimetic representation. Abraham Adams's disquisition on that text, in Book 3 of *Joseph Andrews* (1742), is so detailed that it could practically substitute for the original. Of the Iliad's "action, termed by Aristotle *Pragmaton Systasis*," Adams rhetorically asks, "Is it possible for the Mind of Man to conceive an Idea of such perfect Unity, and at the same time so replete with Greatness?" He goes on to comment on yet more features that give Homer's action its wonderful unity: "As the Subject is Anger, how agreeable is his Action, which is War? From which every Incident arises, and to which every Episode immediately relates."[33] Adams goes on to praise the comprehensiveness and totality of Homer's epic: "If we consider their Variety . . . there is scarce a Character in human Nature untouched in some part or other" (155). Variety was not in fact one of Aristotle's criteria, and it would seem to pull away from the criterion of formal unity. Its invocation by both novelists, then, suggests that there is more at stake in the novel's mimesis—that the novel feels itself to have a special capacity to imitate and to order the diversity of nature's phenomena.

A readily available idea of plot in a literary sense, then, was that of a formal unity. In the well-known preface to *Joseph Andrews*, Fielding draws on Aristotle's literary taxonomy to name what he has written. Aristotle himself had compared an ideal epic plot to the unity of an animal: a plot "should involve a single action, whole and complete in itself . . . so that like one whole living creature it may produce its appropriate pleasure."[34] The magnitude of plot is illustrated by analogy: "A living creature or anything else made up of parts not only must have its parts organized but must also have just the size that properly belongs to it."[35] But the reference to natural variety on the part of both Fielding and Richardson gestures toward a different lineage and different alliances for the novel not only in print culture but elsewhere.[36] The analogy between one animal and one plot is important, but plot now has the task of getting a handle on the existence of empirical variety. It has the capacity not only to display that variety, but by representing it in a single arrangement, to order it.

In addition to formal unity, the inherent organization that was attributed to plot, there was also the deeper uniformity of the laws that regulated the movements of matter in the world the novel represented. These came to lie behind the incidents of the plot in the realist novel. By muthos, Aristotle had meant "the totality of related events . . . as primarily a self-sufficient and self-enclosed causal chain."[37] In the novel, that causal chain of plot contributed to a formal unity, but it also had to refer to the world. That is to say, plot needed to be self-enclosed in a formal sense, but it also had to accept a certain regularity of causation based on the investigation of the physical world. It had to build, encode it, into the way that things happen. In exchange, plot received an entire world of intermediate causes that could be elaborated—causes that could be as tiny as a dropped pocketbook or discarded muff. This dimension could be considered a kind of plot infrastructure: a set of laws that not only governed the imitation of events but was understood to refer in some way to an independently existing world. Through such forms of unity as well as gestures of reference, plot could become a space in which the encounters of different, various natural types could be exhaustively orchestrated, their identities and differences measured, and their ideal relations to one another envisaged. Realist plot, then, proceeds on the assumption of the regularity of nature, unless otherwise indicated. Mechanical philosophy explained how things move or stay still. Natural history showed how things exist not as singularities but as species or types that could be made visible and put into relation with each other.

Within natural history, however, there was disagreement about how to classify the varieties of natural phenomena. The debate between the two highest-profile naturalists of the century, Linnaeus and Buffon, carries some implications for the way that novels use their plots to reveal and relate characters. A major point of contention between them had to do with whether *the process* of comparing and noting a range of differences and similarities between empirical individuals had any epistemological value. The Linnaean system of classifying plants and animals depended on singling out a small number of distinct morphological features and noting their number and position. In the words of one English admirer (and there were many—enough to form a Linnaean Society), Linnaeus was able to establish a system which assigned "the specific names of plants, not taken, as had been customary . . . from the colour of the flower, relative size of the plant, smell, taste, place of growth, time of flowering, name of the discoverer, virtues, uses, duration; none of which are sufficiently permanent; but from those invariable and essential parts, which fully and clearly distinguish each species under the same genus."[38] For plants, these were the number and arrangement of stamens and pistils. The seven orders of mammals were based on the type of feet and "the number, situation, and form, of the three kinds of teeth."[39] On the basis of a selected set of features, all the plants, animals, and insects of the world could be classified according to this "system of nature," as Linnaeus's major work was named. It is worth noting that this system seems to eliminate various orders of time ("time of flowering . . . duration") as the basis of identity.

Buffon, the head of the French scientific establishment and fellow of the Royal Society, pronounced Linnaeus's system arbitrary, devoid of subtlety or depth of understanding. Buffon's *Natural History*, published over several decades, was one of the most widely owned publications in France and, as library auction catalogs suggest, in Britain as well.[40] His influential views of natural order and how it could be ascertained provide a link between Newtonian science and the practices of the realist novel and extend, as Phillip Sloan has argued, to Kant's natural history.[41] Buffon criticizes methods based on abstraction. Though he emulated Newton early in his career and translated Newton's treatise on fluxions into French, Buffon observes that with mathematical sciences, "one is obliged . . . to strip the subject of most of its qualities, and to make of it an abstract entity, and after having arrived at a conclusion equally abstract, when it appears that something real has been found . . . the ideal result is transferred back upon the real subject."[42] This method works for "astronomy and optics," he suggests, but not so well for natural phenomena.[43] What he objects to in Linnaeus's method is a similar formalizing operation: "Wishing to judge the whole by a single part . . . rather than making use of all the parts of the organism, and searching out the differences and similarities of complete individual specimens."[44] The choice of features (teeth, for example), and hence the divisions of nature based on them, are completely arbitrary, he insists. Why not twelve rather than six classes?[45] The alternative approach to classifying nature that he proposes proves to be applicable to the procedures developed in the novel.

Buffon is concerned here with the question of objectivity, calling it "the most delicate and the most important point in the study of the sciences: to know how to distinguish what is really in a subject from what we arbitrarily put there in considering it, to recognize clearly the properties which belong to it and those which we give to it."[46] But the way to achieve it is not through abstraction. Natural history should arrive at its arrangements, Buffon argues, not through a quick enumeration of schematized parts but through a process of intense description and historical characterization. The only means of "advancing natural science is to labor at the description and history of the various things which are its objects."[47] Linnaean naming, he suggests, is a kind of child's play. As opposed to that "dessicated nomenclature," real natural history should "gather together all the objects, compare them . . . extract from the totality of their connections all the insights which may be able to assist us to see them clearly and to know them better. . . . The sole means of constructing an instructive and natural system is to put together whatever is similar and to separate those things which differ. If the individual entities resemble each other exactly . . . such individuals will be of the same species . . . the similarities and differences will be taken not only from one feature but from the whole organism."[48] Description, as Buffon conceives it, possesses an irreducibly narrative character and a novelistic scope, exploring the totality of connections. It should include "form, size, weight, colors, positions of rest and of movement, location of organs, their connections, their shape, their action, and all external functions . . . internal organs," the history "of the entire species; . . . conception; . . . gestation, their birth . . . education, instinct, places where they live . . . customs, . . . services which they can render to us."[49] As he puts it more succinctly, the natural historian's task is "to mark the resemblance; to seize the differences."[50] Here, too, we see an emphasis on laying things out on an imaginary (or real) plane in order to compare them, only in this case the process is iterative and multidimensional, even implicitly narrative in its form.

The method and the goal of natural history as described in Buffon's *Initial Discourse* rest on the legitimacy of representation as a mode of knowing. In this passage, he seems to compare elements as laid out in a static representation; but at the same time, he appears to describe a process of change and movement as it occurs before our eyes:

> It is possible to descend by almost imperceptible degrees from the most perfect of creatures to the most formless matter . . . these imperceptible nuances are the great work of nature, and [man] will find them not only in the size and shape of things, but in the changes, productions, and successions of the whole species . . . nature proceeds by unknown gradations . . . she passes from one species to another, and often from one genus to another, by imperceptible nuances . . . one finds a great number of intermediate species and mixed objects which it is impossible to categorize.[51]

The movement of the eye down a table or page seems to be one with the movement of nature itself, as "she passes from one species to another . . . by imperceptible nuances." He stresses gradations of difference rather than clear-cut types.[52] These gradations occur within species over time as well as between them. Examples of one species can reveal differences that indicate decline or improvement. There is "a general prototype of each species . . . which seems in procreation to be debased or improved according to its circumstances. . . . The original form subsists entire in each individual . . . yet none of these individuals are alike in every thing to other individuals and consequently to the model from which they are formed."[53] Natural history's space of ordering also reflects the relations that natural things have with humans. Buffon begins his natural history of animals with those which are most useful to humans first (the horse, the dog, oxen); then ones that live around him familiarly; then, "We pass little by little to those which are most distant, and which are strange to us."[54] He appears to see no tension between this order's anthropocentrism and its objectivity. Place or placement relative to other creatures, then, expresses meaning. Importantly, though, this space of ordering is apprehended not in a single glance but through repeated, nuanced, iterative encounters. Even the act of looking seems to have a significant duration, for Buffon. But everything can eventually be represented to the eye, given enough time.

Buffon's way of defining a natural fact also has affinities with the protocols of novelistic plot. He emphasizes repetition or duration over time as the condition of factuality. As Lorraine Daston and others have pointed out, the understanding of probability underwent notable development in Buffon's time.[55] In his critique of the mathematical sciences based on "methods of calculus and geometry," Buffon points out that their truths are "only truths of definition," with "nothing of the real about it," whereas "truths of the physical sciences . . . depend only on facts. A sequence of similar facts or . . . a *frequent repetition and an uninterrupted succession of the same occurrences* constitute the essence of this sort of truth . . . a probability so great that it is equivalent to certitude. . . . *It is enough that a thing always happens in the same way for it to become a certainty or a truth for us.*"[56] This statement seems to rely less on the logic of probability than a stubborn empiricism.[57] The recurrence of an event constitutes a fact; its truth lies in the way something happens or comes about. The facts that he has in mind are not those concerning the movement of matter, but rather the existence, among infinite gradations of difference, of natural types that have more similarities with each other than differences: "If the differences between them are so small that they can be perceived only with difficulty, such individuals will be of the same species. If the differences begin to be perceptible, while at the same time there are always many more similarities than differences, such individuals will be of different species, but of the same genus. . . . And likewise this method of inspection will be brought to bear on form, size, external bearing, upon the various parts, upon their number and position."[58] Buffon ends his *Initial Discourse* with the now-familiar refrain that "causes are and always will be unknown to us . . . we must be content to call cause a general effect, and must forego hope of knowing anything beyond that."[59] He adds several new emphases, however, that will prove useful in developing our account of what the plots of realist novels do. He conceives of nature as composed of overlapping gradations of similarities and differences (from which types can be discerned, in and through which types exist). He stresses how regularity of occurrence constitutes fact ("a certainty or truth for us"). He also emphasizes process over system: the activities not just of collecting, but of comparing and describing both specimens and species. Looking or comparing possesses a narrative form in which objective differences appear and disappear over time.

We can locate an intermediate species of writing between natural history and the novel in the "characteristic writing" that Deidre Lynch has examined. Lynch notes that the minor genre of the Theophrastian character underwent a revival in the mid-eighteenth century. In accordance with Foucault's account of an era that located knowledge in "lines, surfaces, forms, and reliefs," Lynch argues that novelists such as Defoe and Fielding "[registered] their culture's investment in the eloquence of the material surface—the face of the page, the outside of the body—and their culture's idealization of what was graphically self-evident."[60] Lists of socially familiar character-types offered readers the pleasures of recognition. One such collection from 1750, "Characterism, or the modern age display'd" (the century's only newly written one, according to Lynch), shows the similarities between the Theophrastian character collection and Buffonian natural history.[61] This anonymous work offers brief portraits of types such as "the hypocritical lady," "the female pedant," "the censorious lady," "the intriguing lady," "the corrupt statesman," "the ambitious clergyman," "the extravagant heir," "the promising gentleman," and many others.[62] Such types could easily emerge out of or retreat back into the pages of novelists from Haywood and Fielding to Trollope and beyond. Yet the format and the descriptions resemble those of Buffon's *Natural History*, with its consecutive brief chapters on the ferret, the weasel, the ermine, the squirrel, the rat, the mouse, and so on.[63] The anonymous author's descriptions of character-types not only make reference to proverbial or emblematic features of animals (the "censorious lady" has a "tongue . . . as venomous as the Sting of an

Adder. . . . She's as blind as a Beetle to the Perfections of her own Sex."[64]), but can be seen to cover in a tongue-in-cheek manner some of the features Buffon recommends to our notice, such as "positions of rest and of movement . . . instinct, places where they live . . . customs . . . services which they can render to us."[65] Of the "promising gentleman, or fashionable friend," we learn that "he never meets you, but he gripes your Hand with as much seeming Sincerity, as if he had a mind to incorporate his own Flesh and Blood with yours, thro' an Affection to your Person. . . . If you happen to rise to any Public Office, or profitable Post, he will stick as close to you upon the Title of friendship as the Ivy does to the Oak, or a Snail to a Cabbage Leaf."[66]

But the form of the catalogue, list, or collection draws our attention to an important difference from novels. Novels must have a plot, even if it's loose or episodic. Description alone cannot make a novel. It is through its plot, too, that the novel has to establish the existence of types and bring them together, relate them to one another, in a way that suggests an objective order. Fielding's interest in types and species is explicitly stated, and his taxonomic logic is clear. Following a familiar method of natural historical classification, he notes that his allegedly new species of writing, the comic epic in prose, has all the parts of epic that Aristotle notes, except one. Still, "It is much . . . more reasonable to give it a Name common with that Species from which it differs only in a single Instance, than to confound it with those which it resembles in no other," he argues in the preface to *Joseph Andrews* (3), following a Buffonian logic. In the same preface, Fielding rejects the genres of burlesque and caricature, both of which show "what is monstrous and unnatural" (4). He aspires only to "the exactest copying of Nature," in which the "Ridiculous" is plainly seen everywhere: "Life every where furnishes an accurate Observer with the Ridiculous" (4).[67] The ridiculous is a natural fact, in Buffon's sense. But I also want to suggest what may be less apparent: that Richardson's *Clarissa* and *Pamela* are equally driven by a taxonomic impulse, though that impulse itself is woven far more deeply into the plot itself. My discussion of these two canonical novelists will be more suggestive than exhaustive. I'd like to think about how their plots perform these activities of placing natural types in a space that is both ideal and empirical, ordering and ordered, a table and a narrative.[68] As Buffon recommends, they "gather together all the objects, compare them . . . extract from the totality of their connections all the insights which may be able to assist us *to see them clearly and to know them better*" (emphasis added). "The totality of their connections" is an excellent description of what plot aims to work out and make visible.

Fielding sets his types in motion in a physical world, makes them encounter each other, and compares the consequences. Sometimes they literally collide with each other; often they pursue one another. Through their physical action and movement, and the characteristic force, velocity, and resistance they offer, such characters reveal their typical nature and establish their relations to one another. Ian Watt offers a striking description of this process: "The plot's function . . . is to perform a physical rather than a chemical change: it acts as a kind of magnet that pulls every individual particle out of the random order brought about by temporal accident and human imperfection and puts them all back into their proper position. The constitution of the particles themselves—the characters—is not modified in the process."[69] The twelfth chapter of *Joseph Andrews* provides a useful example: it's packed with action, reaction, movement, and includes minimal description. Having been fired as footman to Lady Booby, Joseph has just left the second inn on his road back home: "He had not gone above two Miles . . . when he was met by two Fellows in a narrow Lane, and ordered to stand and deliver. He readily gave them all the Money he had." They demand his clothes, but Joseph refuses, whereupon a typical encounter ensues: one robber "levelled a Blow at [Joseph] with his Stick, which Joseph . . . caught with his, and returned the Favour so successfully on his Adversary, that he laid him sprawling at his Feet, and at the same Instant received a Blow from behind with the Butt-end of a Pistol from the other Villain, which felled him to the Ground, and totally deprived him of his Senses. The Thief, who had been knocked down, had now recovered himself; and both together fell to be-labouring poor Joseph with their Sticks . . . threw him into a Ditch, and departed with their Booty" (41). Notice the careful timing of these movements: the instant the first adversary falls down, Joseph receives the blow from behind. The time that it takes for Joseph to lose consciousness seems to be exactly the same time required for the first man to recover himself and stand up again. At this point, a stagecoach containing an old gentleman, a wit, a young lawyer, and a lady comes by, picks up Joseph (after much arguing), and arrives at an inn, whose inhabitants (Mr. Tow-wouse, his wife, and Betty), along with the travelers, all proceed to reveal their typical natures. The lady makes a great hypocritical display of her virtue ("O J-sus, cry'd the Lady, A naked Man! . . . drive on and leave him" [41]), the lawyer and the "Man of Wit" show off their worldliness through a nonstop series of dirty jokes; Mrs. Tow-wouse cares only about money, while the lowly postilion and Betty the maid show their good nature by treating Joseph with kindness.

Unmemorable as this episode might appear, it's important enough for Fielding to allude to it a hundred pages later in the opening chapter of Book 3, in his well-known defense of those authors "who are contented to copy Nature" (147). He is certain that his readers "will know the Lawyer in the Stage-Coach, the Moment they hear his Voice . . . the Wit and the Prude meet with some of their Acquaintance. . . . I declare here once for all, I describe . . . not an Individual but a Species" (148). Fielding notes that such "species" comprise "thousands" of individuals and span human history. They are not, however, "Universals, but would be understood with many Exceptions" (148). Such assertions are compatible with the belief that human species exist as Buffon posits, as patterns of recurrent action or behavior. Here, halfway through his novel, Fielding indulges himself in the clarity of the character-sketch, describing a fixed quantity: "Where extreme Turbulency of Temper, Avarice, and an Insensibility of human Misery, with a Degree of Hypocrisy, have united in a female Composition, Mrs. Tow-wouse was that Woman" (148). But this definitional mode is an exception rather than a rule in his novel.[70] Such modes become somewhat superfluous, only telling us what we've already learned from observing and comparing the actions, movements, and behavior of such characters, who cannot seem to act otherwise than as they do.

Fielding's types are neither ancient nor contemporary; they exist outside history and can be represented in their fullness through such incidents and encounters in the plot. If we look ahead briefly to Walter Scott's novels, we can see the difference. While Scott seems to use natural-historical types in a similar fashion, such types exist only within a concrete historical span. The character of the Baron of Bradwardine, for example, in his 1814 *Waverley*, is distinguished by "the pedantry of the lawyer, superinduced upon the military pride of the soldier."[71] The novel's subtitle, "*'Tis Sixty Years Since*," places Bradwardine in the same decade as *Joseph Andrews*'s publication. But the lawyer and the soldier are not timeless, eternal types as they are in Fielding, nor is their juxtaposition the result of fictive contrivance. Both are created by particular historical circumstances. Bradwardine fought in foreign campaigns because his family's Jacobite politics prevented him from progressing in the legal career for which he had been trained. He is a lawyer-soldier hybrid from a very specific conjuncture that no longer exists. An even more striking example from Scott's novel is found in a Highland chieftain who brings the protagonist into the Jacobite rebellion: "Had Fergus Mac-Ivor lived sixty years sooner than he did, he would . . . have wanted the polished manner and knowledge of the world which he now possessed; and had he lived sixty years later, his ambition and love of rule would have lacked the fuel which his situation now afforded."[72] Urbane and French-educated, he returns to Scotland to play the part of a feudal patriarch. In 1745, it is possible for him to do this, to encourage feudal customs for the sake of his own political ambitions. To his English guest alone, he disparages the enthusiastic poetry of his clan's traditional bard as "the barbarous ritual of our forefathers."[73] Fergus Mac-Ivor could only exist at this historical juncture. He becomes a type when the rebellion collapses. His tragic fate, and the defiance with which he goes to his execution, manage to overcome the dissonance between the French Enlightenment and Highland feudalism, and he acquires a posthumous authenticity. Both characters owe their existence as types to the uneven movement of history, the way that feudal beliefs were only gradually eliminated. History consecrates Mac-Ivor as a type once he has become extinct. Appropriately, or ironically, he ends up as part of a sentimentally cherished portrait in Waverley's home. Both his character and Waverley's become fully visible to the eye in this "large and spirited painting": "The ardent, fiery, and impetuous character of [Mac-Ivor] was finely contrasted with the contemplative, fanciful, and enthusiastic expression of [Waverley]."[74] The double portrait would seem to be an example of taxonomic thinking, as it shows two opposite characters side by side. But Scott's later novel also points to the limits of such a representation, to all that it cannot display, to the lived historical experience that "drew tears" into the eyes of its viewers. Display is now not primarily a means of knowing, but functions as a prompt for private recollection.

However, the plots of both *Pamela* (1740) and *Clarissa* (1747–48) rely on a taxonomic impulse similar to Fielding's.[75] Both are driven by the need to ascertain the relationship between their protagonists and the type or species to which they apparently belong. Are they perfect specimens, or outliers? Are they absolutely singular? Can such a singular being exist in and be part of the natural world without being a monster? Lovelace most often justifies what he does to Clarissa as a test to see whether she belongs to the species "woman," that is, to see whether her behavior under certain highly specific conditions will conform to the predictions that he confidently makes, based on his experience:[76] "Has her virtue ever been *proved*?—Who has dared to try her virtue? . . . To the test then. And I will bring this charming creature to the strictest test that all the sex . . . may see what they ought to be. . . . To the test then. . . . Is then the divine Clarissa Harlowe capable of *loving* a man whom she ought *not* to love?—And is she capable of *affectation*? And is her virtue founded in *pride*?—And if this answer be affirmative, must she not then be a *woman*?" (427–28, original emphases). Lovelace is, of course, the self-styled expert natural historian of "the sex," a concept that names what's at stake in the entire plot. Lovelace even invokes "all the sex" as the audience for whose benefit he performs his terrible experiments. He claims that the entire species of "woman" could be vindicated by the result: it is "the trial of the virtue of her whole sex . . . whether, if once subdued, she will always be subdued" (879). Clarissa is regarded as what Buffon calls "a complete individual specimen," perfect in itself. Yet there is the possibility that she may belong to a class or species that has not yet been discovered and in whose existence Lovelace cannot believe: "Allow me to try if I cannot awaken the *woman* in her?—to try if she, with all that glowing symmetry of parts and that full bloom of vernal graces, by which she attracts every eye, be really inflexible as to the grand article? Let me begin, then . . . and watch her every step to find one sliding one" (431). The botanical metaphors invoke the Linnaean system of classifying species through the disposition of their reproductive organs; indeed, Lovelace's observation of Clarissa's body often focuses on her breasts, which he likes to imagine serving their reproductive purpose.[77] If she stands the test, Lovelace affirms, she'll deserve "from all her sex honours next to divine, for giving a proof that there was once a woman whose virtue no trials, no stratagems, no temptations . . . could overpower" (431). The testing of a woman's virtue is, unfortunately, a venerable plot.[78] What makes Richardson's novel different from earlier iterations is its strange obsession with Clarissa's species-being, and with the epistemological issues involved. Indeed, "Giving proof that there was once a woman whose virtue no trials . . . could overpower" is more likely to overthrow Lovelace's entire system of gender than to confirm anything. Even he seems to suspect that the existence of such a singular woman may belong to the realm of accident and unrepeatable events rather than to the tables of natural history. Even so, the plot relies on a microscopic, relentless collection, curation, and observation (in both sets of letters, Clarissa's and Lovelace's) of every step, every pulse, every flutter of eyelid or hand. And it attempts, tragically, to tie such observational practices to the idea of a nature objectively organized by type. If Clarissa is "divine" (717), her body is also part of the natural world and in that sense has an existence independent of her will.

Both Richardson and Fielding attend to the detailed description of *action* as a series of shifting relations between physical objects located in a single, unified world. The passage above from *Joseph Andrews* illustrates how meticulously Fielding documents the various blows given and received in the fight: their origin, direction, force, deflection (or not). Such description, in his case, obviously alludes to Homeric epic and to the classical epic tradition of describing battles. But in Fielding and also in Richardson, such actions rest on a certain sense of the regularity of physical movements in the actual world. They are not only evidence of personal strength or agility or endurance, as matter continues to move in a straight line unless deflected or stopped, and that straight line is also part of the plot. Movements can thus be predicted because of two sets of reasons. One is based on the sense of an external universe regulated by physical laws. The other derives from this conviction that living creatures are organized into natural types. You can predict what the lawyer will do, just as you can predict how, where, and when a blow will land. And in fact, both Fielding and Richardson spend a great deal of time telling us what would have happened, had something else not occurred to stop or prevent it. This becomes an important dimension of plot objectivity as realized in the novel. It creates a sense of totality by documenting possible trajectories and outcomes that would still have been consistent with these laws of movement and of type.

Several excellent recent studies have examined counterfactuality in narrative.[79] What I'd like to emphasize here, though, is not the divergence from actuality seen in such moments, but rather how they reinforce the idea that movements (or actions) demonstrate a larger objective order. Fistfights and battles in Fielding make this abundantly clear. At the New Inn, a captain "was levelling a Blow at [Adams's] Head, which would probably have silenced the Preacher for ever, had not Joseph in that Instant lifted up a certain huge Stone Pot of the Chamber with one hand . . . and discharged it . . . in the Captain's Face" (201). (The trickling of various fluids down the faces of both men is then meticulously described.) Thirty-some pages later, just as Fanny is about to be assaulted by another minor villain, Joseph, coming up "just as the Ravisher had torn her Handkerchief from her Breast," punches him, and another detailed, almost diagrammatic description follows. Joseph lands the first blow, and delivers the second,

> which had it fallen on that part of the Stomach to which it was directed, would have been probably the last he would have had any Occasion for; but the Ravisher lifting up his Hand, drove the Blow upwards to his Mouth . . . he collected all his Force, and aimed a Blow at Joseph's Breast, which he artfully parry'd with one Fist, so that it lost its Force entirely in Air . . . he darted his Fist so fiercely at his Enemy, that had he not caught it in his Hand . . . it must have tumbled him on the Ground. And now the Ravisher meditated another Blow,

> which he aimed at that part of the Breast where the Heart is lodged, Joseph . . . so prevented its Aim, that it fell directly on his Nose, but with abated Force. (238)

The slow-motion effect of this description is notable. Fielding manages here to narrate movement in such a way as to create the impression of a (confusing) diagram, full of arrows and dotted lines.[80] But all these deflected trajectories are equally lawful. It would require no bending of the laws of motion (or of type) for Joseph's blow to land on his antagonist's stomach, or for the would-be ravisher's to land on Joseph's heart, as they are physical equals. The field of objectivity contains possible events as well as those that are actualized in the plot.

Tom Jones contains similar scenes, most notably upon Tom's arrival at the inn at Upton. Partridge walks into the kitchen exactly at the moment when the landlady's broom is about to descend on Tom's head, and arrests her arm at its highest point. But in this novel, Fielding broadens this way of plotting so that it can include more extended series of motions and events, when a certain object or movement is deflected or intercepted (everything from Allworthy's banknote to Sophia's pocketbook to the movements of Squire Western or Tom himself). Events that could have happened but didn't actually happen are still important to the plot.

This is true of *Clarissa* as well, though somewhat differently. The likely consequences of an imagined act are projected, sometimes carefully calculated. These then feed back into the action that is actually performed—but often with a certain amount of skepticism incorporated into the fabric of the decision. Clarissa fatefully opens Lovelace's first letter to her: "My apprehensions of the consequences of this treatment [by the Harlowes] induced me to *read a letter* he sent me that night" (52). Anna doesn't blame her: "Your condescension has no doubt hitherto prevented great mischiefs" (71). Clarissa takes into account what Lovelace could have, but did not do: "Although provoked by my brother, [he] did not do him all the mischief he could have done him, and which my brother endeavoured to do *him*" (136). An actual parley between the two resembles the dynamic of Fielding's battles: "He was proceeding with reflections of this sort [against the Harlowes]; and I angrily told him I would not permit my father to be reflected upon." Clarissa interrupts, intercepts his blow, here carried out in words. But she then gives him credit for not proceeding in the same line, noting that Lovelace talks about her family "with more coolness and respect . . . than one would have apprehended from a man so much provoked, and of passions so high, and generally thought uncontrollable" (168). Her agreeing to meet him in the garden on the fateful day is based on her calculation of his (and her family's) likely behavior: "If I do *not* [meet him], he may take some violent measures" (263). It is not Clarissa alone who reasons and acts in this way. Her family bases their actions on anticipated behavior and consequences: "This intended violence my aunt often excused by the certain information they pretended to have of some plots or machinations, that were ready to break out from Mr Lovelace: the effects of which were thus cunningly to be frustrated" (348). At this point, Richardson observes in a footnote that "Mr Lovelace artfully contrived to drive [the Harlowes] on by permitting *his* agent and *theirs* to report machinations which he had no intention, nor power, to execute" (348). Shared anticipations have objective consequences.

These realist plots show us characters moving through the world in ways that both reflect certain features of the physical universe and also reveal their own typical natures. In Book 1 of *Joseph Andrews*, Fielding tells the reader to observe "the different Operations of this Passion of Love in the gentle and cultivated Mind of the Lady Booby, from those which it effected in the less polished and coarser Disposition of Mrs. Slipslop" (28). We can already anticipate how this passion will affect in different ways the internal and external movements of their bodies. We can thus locate a natural phenomenon, noting its similarities and differences with neighboring species. These novels do not limit themselves to describing the external features of their characters, though they certainly do that (often, in Fielding's case, in ways that unpleasantly liken women to animals). They set them in movement, to collide with or resist each other: plot serves as a plane of encounter as well as comparison. The taxonomic quality of Fielding's novels has both a synchronic and a diachronic aspect. Synchronically, we get sets of characters whom we're meant to compare and classify: not only unnamed members of species such as "landlady," but also sets that contain members such as Square and Thwackum, who stand related not only as teachers of Tom, but as unworthy beneficiaries of Allworthy's generosity. In the latter category also stand Jenny Jones, Mrs. Miller, and Tom himself. A parallel category would be beneficiaries of Tom's generosity. Within the species "father," the novel invites us to compare and rank not only Squire Western but even Mr. Nightingale's father and uncle, the need for whose existence I can't otherwise explain. One of Fielding's favorite plot formulae, "no sooner had . . . than," stresses that the best way to order these natural phenomena is to look at their instinctive behavior, as Buffon recommends, as well as their relative usefulness to other humans.

Fielding's plots perform for our amusement the activity of scrambling, misplacing, and then replacing and ordering living beings according to the natures that have been revealed through action. The night scene in *Joseph Andrews*, for example, has characters changing locations, going into different rooms, and trying to figure out in the dark who or what someone is through their other senses. Beau Didapper feels and smells Mrs. Slipslop, and vice versa; Adams enters the room, makes his way to the commotion, and, "laying hold of the Beau's Skin . . . and finding his Skin extremely soft, and hearing him in a low Voice begging Slipslop to let him go," believes him to be the young woman in danger, and Slipslop, on whose chin "he found a rough Beard," the attempted rapist (260). Everyone takes the wrong place, just as Joseph and Fanny had been put literally in wrong places, stolen and substituted as infants. Similar things happen at the inn at Upton in *Tom Jones*, when Mrs. Waters's room is continually invaded by various parties looking for someone who could be but is not her. The emphasis on the wrong place does more than show us how amusing (or not-so-amusing) incidents can occur. It speaks indirectly but powerfully of there being such a thing as a right place, a place that shows one's correct relation to all other specimens and species. The ending of *Clarissa*, with its extraordinarily detailed will and last testament, performs the same action of putting things in their correct places, to remain there forever. Realist plots perform this activity of relating and ordering. They rely on representing force, movement, and natural type as lawlike in their occurrence. But they also require a more abstract concept of space as a unitary field of perceptibility charged with meaning and value, where beings can be properly ordered and thus known. The older notion of plot as a single unitary action becomes ideologically powerful in a new way.

I'd like to conclude this chapter by returning to *Clarissa* and the question of taxonomy. I have been arguing that Richardson's plots, like Fielding's, provide a unitary field of visibility in an era when to see was to know, and it was assumed that the important parts or aspects of things could be made visible. We can certainly note the intensity of the scopic drive, the desire on the part of the male characters to see letters, as well as parts of bodies generally hidden, but also the novel's desire to make everything visible, down to each letter that Clarissa will write to be sent after her death. This preoccupation with visibility comes to resemble Buffon's project or method of ascertaining type. The result is a novel—a plot—of unusually extended duration and hypervisibility. The novel's commitment to the latter is clear from the beginning. Anna begins by noting that "every eye is upon you" (41). And Clarissa embraces this responsibility to the very end: "I should be glad that all the world knew my heart. . . . Let them even ask me my most secret thoughts, and . . . I will reveal them" (822). The letters are laid out in full view for us (and the characters) to see. Everything in them is fully visible. But they do acquire a hidden depth when laid out next to one another, when some letters are read as context for another's expression, as Anna warns Clarissa. This dialectic seems inevitable because the novel adopts Buffon's standard of factuality: recurrence, or "a frequent repetition and an uninterrupted succession of the same occurrences." Lovelace writes, "Nor is one effort, one trial, to be sufficient. Why? Because a woman's heart may be at one time *adamant*, at another *wax*—as I have often experienced" (430). Such knowledge is one of the "libertine maxims" (430) that Lovelace wishes both to confirm and to refute through his endless plots. One hears the echo of Buffon: "It is enough that a thing always happens in the same way for it to become a certainty or a truth for us." But how many trials are sufficient? Which instances serve to reveal, and which to obscure? "If the differences begin to be perceptible," Buffon writes, then it may be a case of different species. But when can a difference be said to begin?

Both Lovelace and Clarissa seek to know (or refute) their species-being. Belonging to a species is proven by recurrent behavior over time, not by the number or static arrangement of certain parts. Lovelace confidently predicts to Belford about Clarissa, "Modesty of the sex will stand my friend at any time. At the very altar, our hands joined, I'd engage to make this proud beauty leave the parson and me, and all my friends present . . . and this only by a very word" (602). And he is not wrong. When he mentions settlements to her, she can't accept them. But this is due not to instinctive behavior on her part, but to her inexplicable commitment to a standard of consistency. "Were not hesitation, a self-felt glow, a downcast eye, more than enough?" she writes to Anna. "I thought myself obliged . . . on the one hand, to save myself from the mortification of appearing too ready in my compliance, after such a distance as had been between us; and on the other, to avoid . . . the necessity of giving him such a repulse as might again through us out of the course" (594). The virtuous woman operates through gradually, infinitesimally diminishing microrepulses. The libertine operates through increasingly outlandish sequences of aggressive movements. Clarissa and Lovelace are equally constrained by type, as much as they both wish to transcend it and be recognized as absolute and unconditioned in their being.

Clarissa's plot, in its broadest scope and purpose, seems unable to escape the logic of type, in which actions are viewed as either consistent or inconsistent, predictable or unpredictable. Lovelace on Clarissa: "So . . . gloriously consistent in the character she has sustained . . . to the present hour" (734–35). Clarissa on Lovelace: "Has he not given you a flagrant specimen of what a man he is?" (799). What is remarkable, though, is the degree to which both characters yearn to escape both type and plot itself. Both regard prior events and actions—especially their own—as terrible constraints, inescapable conditions of any future act. Lovelace laments, "I cannot bear, even to be *compelled* to be the wretch my choice has made me . . . it is a very foolish thing for a man . . . to have brought himself to such a height of iniquity, that he must proceed, and cannot help himself" (848). In the end, the principle of plot objectivity alone seems to triumph. Both Clarissa and Lovelace have wrongly tried to believe that plot consists in choice alone, that the power of choice resides in some inner reserve, or that character can, at any moment in the plot, transcend that tangle of iteration. There is no more ironic statement in the entire novel than the one that Clarissa writes about meeting Lovelace in the garden: "What he requires . . . cannot affect my future choice and determination" (263). In this novel, even more than in others, each step lays down the conditions of the next one; in Lovelace's words, "What is done, is done" (411). At the very end, Clarissa seems to acknowledge this. She blames her fate on a "strange concurrence of unhappy causes" (1301), on the totality of connections that is the plot itself, laid out before our eyes.

CHAPTER 5

Tracing Change and Testing Substances

Intimate Objectivity

In Ann Radcliffe's novels, prisons seem to have a strange habit of opening themselves. In *The Italian* (1797), for example, we are trapped in a subterranean vault: "The door closed, with a thundering clap that echoed through all the vaults." It appears impenetrable. "The thick wood was inlaid with solid bars of iron; and was of such unconquerable strength, that it evidently guarded what had been designed for a prison."[1] We can imagine that Robinson Crusoe would struggle to open (or maybe to build) that door for years. Yet one chapter and a few hours later, the door "stood a little open": Vivaldi "could scarcely believe his senses, since . . . he had not heard its ponderous bolts undrawn" (99). It is not possible for the door to open itself, and yet it's never explained who opened it. Emily, the heroine of *The Mysteries of Udolpho* (1794), is pressured by Montoni to sign over the rights to her estates in France; this seems to be the reason he holds her captive in Udolpho after her aunt's death. She eventually signs. He does not fulfill his promise to release her, but then he never takes possession of the estates either. Her release, like his later death, comes about unexpectedly and randomly. After a duel between two people but not as a result of it, the servant Ludovico appears in her room and orders Emily to follow him. "Emily enquired what had occurred, and whither they were to go? 'I cannot stay to tell you now, Signora,' replied Ludovico: 'fly! fly!'"[2] A gate had been left open. But what created this opportunity? Apparently, it has less to do with an act or event than with the composition of the castle itself: solid, massive, and imposing on first sight, Udolpho turns out to be porous, crumbling. It is full of holes, passages, dead ends, and openings, not unlike the plot itself.

In the novels examined in this chapter, actions can seem curiously underdetermined. Their plots seem less interested in questions of agency, of who made what happen, than in questions of composition, of how something behaves because of what kind of substance it is. They are more absorbed in *tracing change* than in *recounting actions* and their consequences. We are less likely to find someone making, carving, planting or even moving things in these novels than entities being moved, being conjoined with or separated from other entities by means not always emphasized. They will be closely observed in various situations and from various distances. This is true not only of Radcliffe's novels, where the characters can appear subordinate to their surroundings like figures in a landscape painting, but also of Frances Burney's. Burney's plots typically place their female protagonists in situations of extreme constraint, pressure, or duress, where they are forced to act in ways that do not reflect who or what they are. Her novels seem to pare actions and agency away from something like the objective substance of character. It is almost axiomatic in Burney that for the protagonists, what I did is not in fact who or what I am. This can even identify them as protagonists. Think of the rude cousins in her first novel, *Evelina* (1778), for instance, or the many minor characters who act, unauthorized, in the protagonist's name. They grab things, buy things, borrow money, all in her name. Or the protagonist, under duress, does something that contradicts who she is, and of course someone important happens to be there to see it. Clearly, social identity matters in Burney's plots, but that's not the only dimension of identity that stands to be revealed. Identity reaches as deep as the question of what, not just who someone is, of material substance rather than name or social class, because a certain kind of objectivity still remains at the heart of plot.

It is notable that self-consciousness is not given more weight in establishing personal identity; in terms of the plot, self-consciousness is surprisingly ineffectual. Even when agonizingly intense, a subjective sense of identity is not sufficient *for the purposes of plot* to constitute the truth of a character or the reality of its world. What I feel or know myself to be is not the crux of the matter. Someone or something else needs to observe, to confirm what I am made of. And plot is given this task. It sets up, defers, complicates, defeats, and ultimately offers this certainty. The protagonist's character cannot be deduced from her acts alone, much less the consequences that ensue from them. It is conceived in terms of the qualities of the substances, or the stuff of which *they and their worlds* are made: how porous or elastic or transparent or ductile? How reactive, how volatile? Such questions are answered by observing how a material substance behaves under various conditions, how it reacts and changes or doesn't change. These plots exhibit a need to test and test again, to dissolve its characters into their world, to see or to show the stuff they are made of. In other words, objectivity is sought not only in a world external or resistant to characters' aims, or in an ordered array of natural types. In the realm of plot and for the purposes of plot, objectivity extends to the substance and inner composition or contexture of the represented world and the bodies in it. Plot also posits dynamic natural forces that operate at a different scale of intimacy and in a different manner from bodies that move or push against each other.

Mechanical philosophy, as seen in the following statement, tended to handle questions of composition in terms of the spatial arrangement of identical units: "Matter is the same in all Bodies . . . the whole Variety of Bodies, and the different Changes that happen in them, entirely depend upon the Situation, Distance, Magnitude, Figure, Structure, and Cohesion of the Parts that compound them."[3] But this belief was supplemented by another way of examining changes that occur in bodies. Consider this description, for example:

> The green glass became first of a bluish colour on the surface, and in this state . . . it appeared less transparent than before. . . . After this it was found changed a little way on both sides into a white substance . . . as this change advanced further and further within the glass, the vitreous part in the middle approached more and more to yellow: the white coat was of a fine fibrous texture, and the fibres disposed, nearly parallel to one another. . . . By degrees, the glass became throughout white and fibrous, the external bluishness at the same time going off, and being succeeded by a dull whitish or dun colour. . . . The pieces which were continued in the fire for any considerable time beyond this period . . . suffered a fresh change, which proceeded, like the first, from the surface to the center. The fibres became divided or cut into grains at the outer ends, and by degrees they were thus successively divided through their whole length; the whole internal part of the porcelain assuming a granulated texture. . . . Those which were longer and longer exposed to the fire, received more and more alterations. The grains, at first fine and of some degree of glossiness, grew larger and duller; and at length, through sundry gradations . . . what had once been glass, and afterwards a compact hard porcelain, became a porous friable substance, like a mass of white sand slightly cohering.[4]

This passage comes from William Lewis's *Commercium philosophico-technicum; or, the philosophical commerce of arts: designed as an attempt to improve arts, trades, and manufactures* (1763–65). While Lewis, an experimental chemist and fellow of the Royal Society, was later recognized for his work in titrimetry, what may strike a reader of novels is the remarkable detail found in his description of qualitative change. The changes of color, opacity, and texture that occur in this heated piece of glass over time are noted, as well as the vectors and gradients along which such changes occur ("from the surface to the center"). No quantities are measured or noted, which distances this work from the usual narrative of the "chemical revolution," or the advances made in the eighteenth century through precise quantitative measurement. Lewis does not even measure the time the process takes. His description seems entirely absorbed in its endeavor "to trace the gradual progress of the change from the state of glass to that of perfect porcelain and to discover whether a continuance of the process would be productive of any further changes."[5] That is not a bad description of plot if you imagine characters as the material that changes gradually from one state to another. Lewis's meticulous description dwells in its own sense of time, its own duration. It is not simply that he relates a narrative. How he tells the story entails noting an incremental process of change: the objective, time-dependent transformation of a body—of a substance—under certain conditions or within certain parameters. What is being scrutinized is the process of change undergone by this body. Obviously, the body is not a living, organic, or sentient one. It is a piece of ordinary green glass being heated. But to refer to these changes as mechanical does not fully capture the sense of the process. The successive and gradual changes in color, opacity, and texture that occur when the glass is subjected to intense heat (again, not quantified) do not seem to exhibit an external force acting on a thing mechanically. They seem rather to show what this substance is *through what it becomes, and how it endures*. The transformation from glass to porcelain and beyond possesses an intimacy that feels different from the movement of bodies through space. And the emphasis shifts from an external causality (who or what caused the change) to a process of becoming, observed under certain conditions.

In the novels of the later eighteenth century that I will examine here, there seems to be a change in scale of the deep plot's orientation. It shifts toward the small, the intimate, almost the subsensible. These novels still include large-scale visible types in order to rank and relate them—particularly in Smollett, as we will see. They also answer questions of who or what caused something to happen, but notoriously badly, in the case of *Udolpho*, where the cause of a mysterious event or phenomenon too often turns out to be smugglers. But there is more interest in qualitative change as an event, and especially in the way that changes occur *within or between* bodies as a result of their inherent qualities, their proximity, or other conditions. Instead of a push or a blow, one body moving or resisting another, we see change happening at a more intimate and yet objective level. Plots will try to trace gradations of change in a character, a sound, or even the shifting colors in a sunset. But can the description of a changing sky count as plot? Shouldn't it be assigned some other function in a novel? Isn't it simply setting the stage or providing the background for something that will happen, or perhaps indexing the existence of the narrator or some thematic preoccupation? Without denying that description can serve these other functions, this chapter suggests that these novels do not make a sharp distinction between plot and the description of change, between the observer and the observed, or even between human character and material setting or surrounding. The latter distinction in particular becomes particularly fragile. These novels disclose a world in which bodies are porous and forces are pervasive. It turns out that many descriptions in these novels are, like Lewis's account of heating glass, attempts to capture some kind of transition occurring at a deep level in what's being observed. Sometimes it extends over a long period of time; sometimes such transformations are concentrated in a moment, in a gesture or spasm. In either case, qualitative transformations hold out a clue to what something or someone is, or could become.

This chapter will relate such interest in qualitative change to the emergence of chemistry in the eighteenth century as a branch of natural philosophy. Without rejecting mechanical explanation but without relying on older paradigms, whether alchemical or Aristotelian, eighteenth-century chemists sought to describe why and how different substances reacted to each other the way they did, why some were attracted to others more or less strongly, and what the distinctive nature of chemical combination or separation might be. The chemical processes in question were not limited to those that were performed in a laboratory. As Alistair Duncan points out, they included "burning, rotting, rusting, and many other natural processes" that are familiar and yet strange: "They involve the substance that burns or rots or rusts changing into something completely different."[6] Such phenomena could not be fully explained in terms of a mechanical model of causality, though the latter was not rejected. In order to justify its status as an experimental philosophy and to guide its practice, chemistry had to develop "new kinds of explanation . . . the conceptual model of chemical composition and chemical change, different from the conceptual models of the mathematicians and physicists."[7] Mixture, for example, was a phenomenon that seemed to require different pathways of investigation and different models of explanation, even if those findings also took the form of tables and classificatory schemes. Investigating composition involved more than just collecting specimens or observing them. If natural history aimed at the "observation and collection of things 'given by nature,'" chemists manipulated their materials much more actively.[8] In parallel, I want to suggest, the plots of some influential and popular novels test and observe the intimate composition of substances as revealed through their qualitative reactions. Note that the goal here is not to ascertain whether something belongs to a certain species, or how far it conforms to a prototype. Change is conceived in an inside-out and mediated way rather than as an act performed on a body from the outside or through a mere change in location or situation. It is attributed to qualities inherent in the substances themselves rather than to a temporally delimited act or event. Is the change superficial or deep? Is it temporary or permanent? Can it be undone, and if so, how? Plots pose these questions again and again even within the same novel. This holds true as well of Jane Austen's novels, for example, which will be the subject of the next chapter. Each of her plots reflects on the difference between permanent change, reversible change, and nonreactivity.[9] I am arguing not that novels' plots imitated science or vice versa, but rather that we can see a concordance at a certain level between the questions that both asked about the world and how things happen in it.

Though the discoveries of Antoine Lavoisier that were often summed up as the chemical revolution drew much attention from historians of science in the twentieth century, recent studies of eighteenth-century chemistry argue for important developments that encompass more than the work of this single figure.[10] Chemistry occupied an ambiguous, elusive, and changing position in the field of natural philosophy; before the middle of the century, it was seen as a minor branch of science, tied on the one hand to alchemy and premodern forms of knowledge, and on the other to artisanal practice and medicine. As Ursula Klein argues, it was preoccupied with materials that it conceived and investigated in multiple ways rather than a single one. Chemical substances were "not unambiguously constituted as philosophical objects, as were the atom, the vacuum, the magnetic and electrical fluids . . . did not fit exclusively into the category of perceptible natural objects, such as plants, animals, and minerals . . . [were] not exhaustively defined as useful materials applied in the arts and crafts and the wider society."[11] The substances in question were all of these, and their properties and uses were constantly being debated and manipulated. Klein notes that until the late eighteenth century, chemists did not "[organize] their experiments into a systematic investigative pathway"; they frequently would test a substance with different reagents and solvents, the number of which increased in this period, and report the results.[12] Chemistry worked as something of a bridge between natural history, which focused on "visible objects presented to our senses by nature" and experimental philosophy, which investigated "underlying structures, processes, and causes."[13] Eighteenth-century chemists "moved from the perceptible to the imperceptible dimension of substances, and vice versa."[14] A crucial problem was that of the nature of a mixture or compound. In the alchemical view, a compound was "a homogeneous union which did not preserve the original constituting substances." But eighteenth-century chemists came to believe that a "chemical compound meant a substance consisting of different substance components held together by chemical affinities . . . in a true chemical transformation . . . the original substances were preserved . . . and held together by chemical affinities."[15] Rather than resulting from an irreversible transformation, compounds could be undone, these chemists insisted, under certain conditions: this made it "a third way between transmutation and mechanical mixture."[16]

Peter Hanns Reill argues that important changes in conceptualization started around midcentury. Where chemical phenomena had previously been interpreted in terms of the laws of motion and attraction, they began to be understood differently. "Relations between active or activated material substances replaced the action and reaction of abstract forces. In this theoretical and practical reorientation of natural philosophy the central explanatory features and most of the lexical items of the mechanical philosophy of nature were dropped, ignored, or . . . drastically redefined. A new language of nature was being formed . . . the 'compositional revolution' in chemistry."[17] The change was not limited to systems of nomenclature or even experimental practices, Reill argues. These shifts in the latter half of the century "incorporated the desire to resubstantialize the universe." Instead of Newton's vision of "a world consisting of vast, empty spaces infrequently inhabited by tiny specks of hard impenetrable matter held in place by invisible forces acting at a distance, the newly substantialized world of the late eighteenth century was completely filled by material substances. They ranged from immediately perceptible substances to a host of . . . subtle fluids."[18] By the early 1760s, William Lewis was able to articulate succinctly how the two sciences of chemistry and mechanical philosophy proceeded from different ontological assumptions. Both investigate "the properties of bodies," he asserts. However,

> Natural or mechanical philosophy seems to consider bodies chiefly as being entire aggregates or masses; as being divisible into parts, each of the same general properties with the whole . . . gravitating, moving, resisting, &c. with determinate forces, subject to mechanic laws, and reducible to mathematical calculation. Chemistry considers bodies as being composed of such a particular species of matter; dissoluble, liquefiable, vitrescible, combustible, fermentable, &c. impregnated with colour, smell, taste, &c. or consisting of dissimilar parts, which may be separated from one another, or transferred into other bodies. The properties of this kind are not subject to any known mechanism, and seem to be governed by laws of another order.[19]

If mechanical philosophy ponders how much force it would take to move something, chemistry asks other questions: Can it dissolve? Does it burn? How does its color change? How does a substance change its qualities entirely when combined with another? Lewis argues that mechanical laws cannot provide answers to such questions.[20] Chemistry sees bodies differently. A body may reveal an internal heterogeneity under certain conditions; it may be composed of parts that are not all alike or consistent. Of particular note is Lewis's account of elective affinity, which became a core concept for chemical investigation.

> To the grand active power, called *attraction*, in the mechanical philosophy, what corresponds in the chemical is . . . *affinity*. The mechanical attraction obtains between bodies considered each as one whole. . . . When the attracting bodies have come into the closest contact we conceive, they still continue two distinct bodies, cohering only superficially, and separable by a determinate mechanic force. The chemical attraction, or affinity, obtains between bodies as being composed of parts, and as being of a different species of matter from one another. It never takes place while the two bodies are at any sensible distance; and when they are brought into the closest contact, there is frequently necessary some other power, as fire, to excite their action

> upon one another. In proportion as this action happens, they are no longer two bodies, but one; the affinity consisting in the intimate coalition of the parts of one body with those of the other. The properties of this new compound are not in any kind of ratio of those of the compounding bodies. . . . As the chemical union, and the properties thence resulting, are exempt from all known mechanism, so neither can the bodies be separated again by mechanic force. But a third body may have a stronger affinity to either one of the component matters than they have to one another, in which case, on presenting to the compound this third body, the former union is broken, and one of the first bodies coalesces with the third, while the other is detached and separated. (iv–v)

Lewis is not presenting an original account of chemistry but summarizing generally accepted views. Several points are worth noting for their implications with regard to novels and their plots. In the view of mechanical philosophy, as Lewis presents it here, bodies are essentially homogeneous; they can be divided by mechanical means, and each of the pieces will have the same properties as the original mass (thus, a body is an aggregate). These are Newtonian bodies, subject to laws governing their movement as integral entities. But in Lewis's chemical view, bodies consist of parts that collectively give them their distinctive perceptible properties. When brought into contact, chemistry's bodies do not simply impart force or motion to each other, or attract or repel each other, but actually combine into something new through an "intimate coalition." In comparison to such a bond, mechanical attraction comes to appear as a superficial relation between "two distinct bodies." In a "chemical union," though, two bodies brought together in close proximity and subjected to some form of excitement, as he puts it, actually become a single thing. Lewis asserts that external or "mechanic force" cannot separate the parts of such a composition, no matter how strong it might be. But that body can be separated or decomposed if some other substance is added with which an original component has a stronger affinity, in eighteenth-century terms.

The different affinities that substances have with each other had been placed into a table toward the beginning of the century by the French chemist Etienne-François Geoffroy. Geoffroy drew up a table the order of whose columns and rows indicated the relative strength of the "rapport" that various elements had with each other: in other words, how elements of one compound would leave it and unite with another, in the way that Lewis describes above.[21]

The social analogy that immediately suggests itself had precedents in the way that earlier forms of science, including Greek philosophy, had thought about chemical reactions; for alchemists, for example, "substances involved were personified and moved by emotions of love and hate."[22] The term, "affinity," applies the metaphor of kinship, but can obviously have other implications. What we might call the plot possibilities of such a table of affinity were not lost on the natural philosopher Fontenelle, who wrote in 1718

> that a body which is united to another . . . should leave it to go and unite with another which is presented to it, is a thing of which the possibility would not have been guessed by the most subtle philosophers, and of which the explanation is still not too easy for them today . . . all the experiments of Chemistry prove that a particular Substance has more disposition to unite with one Substance than another, and that this disposition has different degrees. . . . This Table becomes in some sort prophetic . . . it can foretell the effect and result of the mixture, because one can see from their different relations what ought to be, so to speak, the issue of the combat. . . . A Chemical Table is by itself a spectacle agreeable to the Spirit.[23]

As Reill observes, "Many eighteenth-century thinkers assumed that the ties of chemical attraction were analogous to those of human sympathy . . . the language of human relations and chemical reactions were easily interchanged."[24] The analogy to human behavior seemed irresistible. In her *Conversations on Chemistry* (1806), Jane Marcet, the author of numerous works of science education, explains this concept in the form of a dialogue between a teacher and her two young female students. One pupil remarks with regard to affinities, "We might, I think, use the comparison of two friends, who were very happy in each other's society, till a third disunited them by the preference which one of them gave to the new-comer."[25]

Goethe, of course, developed the novelistic potential of this analogy in his strange and fascinating 1809 novel, *Elective Affinities*, whose plot traces how a marriage is broken apart by the incorporation of two new persons into the household, and how new bonds develop as a result. The characters themselves discuss the chemical analogy in a conversation early in the novel, in which two versions of the basic plot are offered. One is schematic, using language similar to Lewis's: "Imagine an A closely bound to a B and by a variety of means and even by force not able to be separated from it; imagine a C in a similar relationship with a D; now bring the two pairs into contact; A will go over to D, C to B, without our being able to say who first left the other, who first with another was united again."[26] The other description emphasizes the role of a human observer and suffuses the process with pathos: "These entities . . . need an observer who will watch with some engagement of his sympathy how they seek one another out, how they attract and seize, destroy, devour, and consume one another and at once emerge from the closest possible union in a renewed and novel and unexpected form."[27] The second description brings out the full dramatic potential of this chemical reaction; it prepares us for the plot that will follow, and the extraordinary, inexplicable intensity of the new attractions that will be felt by the characters.[28]

This chapter, however, will not focus on the concept of elective affinity as an abstract formal template for plot. Rather, I want to show more broadly how, in some novels, plot investigates the composition of characters and the world they inhabit by looking at how they change over time and affect each other. Chemistry was understood to observe natural phenomena in this way and at this intimate scale. Marcet's teacher in *Conversations on Chemistry* begins by explaining to her pupils, "Chemistry . . . is so closely connected with natural philosophy, that the study of the one must be incomplete without some knowledge of the other; for, it is obvious that we can derive but a very imperfect idea of bodies from the study of the general laws by which they are governed, if we remain totally ignorant of *their intimate nature*."[29] She goes on to quote the late eighteenth-century Swedish chemist Torbern Olaf Bergman, who places chemistry highest and last in the progress of the knowledge of nature: "The first [period] was that in which the attention of men was occupied in learning the external forms and characters of objects, and this is called Natural History. In the second, they considered the effects of bodies acting on each other by their mechanical power, as their weight and motion, and this constitutes the science of Natural Philosophy. The third period is that in which the properties and mutual action of the elementary parts of bodies was investigated. This last is the science of CHEMISTRY."[30] Again, motion is distinguished from genuine change, the latter being the result of "mutual action of the elementary parts of bodies." Marcet's teacher states, "The object of chemistry is to obtain a knowledge of the *intimate nature of bodies, and of their mutual action on each other*."[31] I want to suggest that plots in Burney and Radcliffe think in a different way about what counts as objective change or identity. Instead of resulting from the application of force, or from the manifestation of natural type, change occurs (or fails to occur) within something conceived as a compound substance. The substance in question is guessed from its manner of reacting to things, the qualitative changes it exhibits. It reaches beyond moral character or moral qualities to something deeper, more intimate, and yet coextensive with the external world, not sharply demarcated from it. To think of character as substance is not to set it in opposition to soul or spirit, or to imply that humans are machines. Indeed, the chemical analogy rejects mechanism. Rather, I want to suggest that this intimate substance of character is part of the ontological repertoire of plot. Here, plot seems most invested in marking the intimate, qualitative changes that ensue when characters are subjected to various kinds of pressures and proximities.

We can find something of a transitional example in Tobias Smollett's final novel, *Humphry Clinker* (1771). Written as a series of parallel correspondences, the novel not only incorporates familiar social-natural-historical types, but uses them as organizational principles. There is Matthew Bramble, the cantankerous but soft-hearted old man whose letters provide the anchor of the novel, a romantic young girl (his niece), an older unmarried woman (his sister), and so on. Each writes to a chosen correspondent in a manner that reflects their fixed preoccupations. As these characters travel together for health or amusement through Britain and come home again, various accidents and encounters occur; an illegitimate son is discovered, several marriages are contracted, and a degree of health restored to the hypochondriac Bramble. What makes this novel relevant to this chapter, however, is its strange obsession with mixture as a phenomenon. The many mixtures that are vividly and disgustingly detailed make a polemical case against modernity: the real fermenting and amalgamating substance flowing through Britain is, of course, money.[32] But the way that Smollett accounts for mixtures and compositions situates the novel's ontology between mechanical philosophy and chemistry, but closer to the former. It mostly reflects mechanical force, or the attraction, movement, and separation of solid bodies rather than chemical affinities or more intimate processes of change. Like Smollett's earlier *History of an Atom*, which traces the changing location of its narrator (a single atom), *Humphry Clinker*'s idea of change consists largely of the movement, arrangement, or collision of particles. It shows that no mixture defies analysis. But most mixtures can ultimately be separated *by mechanical means*, by the application of force, and its components can be put in their right places.

Unsurprisingly, given his medical background, Smollett's familiarity with chemistry and chemical terms is apparent: nearly every fluid or atmosphere that the characters encounter in their travels is analyzed by Bramble as a composition. The water in Bath is described as "a delicate beveridge [*sic*] . . . medicated with the sweat and dirt . . . and the abominable discharges of various kinds, from twenty different diseased bodies, parboiling in the kettle below . . . the straining of rotten bones and carcasses."[33] The air at the assembly is "a compound of villainous smells . . . essence of mingled odors, arising from . . . sweaty feet, running sores and issues, plasters, ointments, and embrocations . . . besides a thousand frowzy steams" (76). (I leave out the even more nauseating ingredients.) In London, the water is even more troublingly complex, "composed of all the drugs, minerals, and poisons, used in mechanics and manufacture . . . mixed with the scourings of all the wash-tubs, kennels, and common sewers, within the bills of mortality" (135). The wine, the bread ("a deleterious paste, mixed up with chalk, alum, and bone-ashes" [136]), and especially the milk of London are all likewise adulterated: "The milk itself should not pass unanalysed, the produce of faded cabbage-leaves and sour draff, lowered with hot water, frothed with bruised snails, carried through the streets in open pails, exposed to foul rinsings, discharged from doors and windows, spittle, snot, and tobacco-quids from foot passengers, overflowings from mud carts, spatterings from coach wheels, dirt and trash chucked into it by roguish boys" (137). Smollett's novel is obsessed with the idea of purity, which is conservatively identified with remaining within the spatial and social boundaries of one's origin—or, better yet, of one's own property. Matthew Bramble, whose letters provide the above descriptions, frequently extols the purity and excellence of everything produced and immediately consumed on his own estate. The source of all the contamination he notes in his travels is what he portrays as a universal mania for motion, as in his description of London: "All is tumult and hurry. . . . The foot-passengers run along as if they were pursued. . . . People . . . drive through the streets at full speed . . . the whole nation seems to be running out of their wits" (101). London milk clearly has problematic origins (i.e., cabbage and snails rather than a cow), but its worst impurities come about because the milk itself travels through the streets "in open pails," and receives the physical matter ejected from the vehicles and people speeding along beside it.

Yet there is a sense that the nauseating mixtures and "compositions" that fill these pages are "separable by a determinate mechanic force," in Lewis's words. They could have their worst contaminants removed by mechanical means—by straining and filtering, for example (though you still might not want to drink it), by relocating a pump or by scrubbing out a cistern. Order can be restored through attention to such things, by scrubbing off dirt, and building up proper walls and boundaries, all of which happen in the course of the plot more than once. The first time he is seen, Clinker is a ragged, "shabby wretch" with torn clothing; when he puts on a clean uniform, he is described as "metamorphosed" (95), but the change is not even skin deep. Even when seen in the torn clothes of a "beggarly rascal," his skin is described as "fair as alabaster" (93), following the usual romance trope. The plot, such as it is, culminates in an accident when the coach overturns in a river and Bramble nearly drowns. Maximum disorder gives way to the restoration of order, as mistaken identities or disguises are then revealed in comically rapid succession. Immediately after saving Matthew Bramble's life, Clinker is discovered to be Bramble's illegitimate son. "Almost at the same instant," his niece Lydia writes, "a gentleman, who came to offer us his assistance, and invite us to his house, turned out to be a very old friend of Mr Bramble" (372). The disreputable actor who had been courting Lydia turns out to be this highly respectable gentleman's son in disguise. When their identities and origins are thus mechanically revealed by the plot, order is restored. Things and people are put back in their proper places, like with like and unlike or unfit properly segregated or removed. The final episode of the plot, involving a friend of Bramble's, repeats an earlier story of an estate being restored to order by ejecting (selling off, laying off) what is extraneous or impure, using mechanical force to detach or remove impurities. Order is a matter of putting things in their proper place, and not allowing them to move too far. Though one or two characters receive a degree of polish purely as a result of the movement and friction they have undergone, everyone remains exactly what they always were, of which there was never any doubt. There is practically no "mutual action." Change occurs not in the internal substance but in the external disposition of unchanging things.

But much of the anxiety palpable throughout Smollett's novel seems to come from its awareness of the porousness of bodies.[34] Impurities can be deeply absorbed, and who knows what changes might result? The novel's dominant thematic vocabulary of disease and health, corruption and purity, allows it to circumvent deeper uncertainties about change. At the novel's end, Bramble acknowledges that his health has improved as a result of his travels, and opines that "we should sometimes increase the motion of the machine, to unclog the wheels of life. . . . I have even found a change of company as necessary as a change of air, to promote a vigorous circulation of the spirits, which is the very essence and criterion of good health" (377). The model is still pneumatic rather than chemical, a question of rate of flow. And the novel's desire is for a complete, static, and unitary order of the kind that Bramble envisions when he goes to visit the British Library: "I could wish . . . the whole of the animal, vegetable, and mineral kingdoms completed, by adding to each, at the public expence, those articles that are wanting. It would likewise be a great improvement, with respect to the library, if the deficiencies were made up, by purchasing all the books of character that are not to be found already in the collection—They might be classed in centuries . . . and catalogues printed of them and the manuscripts." Bramble also expresses a wish "that there was a complete apparatus [in the library] for a course of mathematics, mechanics, and experimental philosophy" (116).

Despite its attention to bodily fluids and the exaggerated idiolects of its epistolary voices, Smollett's novel offers at most a sense of pseudointimacy. It embraces an epistolary form that already feels worn out, and relies on venerable characterological stereotypes. We can also cite the novel's conception of composition as essentially aggregative. Its political message rests on the hope that mechanical force, applied from the outside, can separate the components of these adulterated compounds and put things back in their places, or at least regulate their movements. Frances Burney's novels, in contrast, create an almost agonizing sense of intimacy. This has to do with more, I think, than their representation of consciousness.[35] Rather, the plots of her novels bring to mind Marcet's definition of chemistry: "The object of chemistry is to obtain a knowledge of the intimate nature of bodies, and of their mutual action on each other."[36] Her plots work to create such knowledge in a dialectical manner. To put it paradoxically, Burney's characters cannot be what they are; and they can be neither separated from nor fully united with one another. Plot's function is to show these unfortunate truths. Certainly, characters struggle to reveal who they are through their actions; but beyond that, Burney's plots suggest an intimate substance that must be tested but cannot be made fully visible to others.[37] By plot here, I refer both to the sequences of events and the ontology that underlies them. My argument will focus on Burney's third novel, *Camilla* (1796), in which plot necessarily fails to disclose the inner, objective substance of the characters of Camilla and Edgar, even as it traces the mutual action of these bodies on each other, their incomplete combination and incomplete separation.

Camilla's plot begins with an elaborate apparatus through which the destinies of a set of characters appear to be planned out by Camilla's well-meaning but disastrously inept uncle. While Camilla's counterpart, Edgar, is supposed to marry another character, that plan soon turns out to be insignificant as an obstacle. The mutual attraction of Camilla and Edgar is clearly manifested and parentally approved from the beginning of the novel. Early on, Camilla's parents "witnessed the involuntary movements which betrayed their mutual regard with the tenderest satisfaction."[38] (Note the "involuntary movements.") What then keeps the lovers apart for seven more books and nearly seven hundred pages? Neither misfortunes nor accidental separations but rather what Burney describes as "mutual solicitude, mutual watchfulness, and mutual trials of each other's hearts" (679). Camilla and Edgar cannot be joined, but neither can they bear to be separated. Their bond cannot be broken, and yet they cannot manage to feel assured of what the other is. After becoming engaged about midway through the novel, Camilla dramatically (and quite understandably) decides that she cannot take any more of Edgar's continual doubting, testing, and watching. She dissolves their engagement, proclaiming, probably to the relief of many readers, "I here, therefore, solemnly release you from all tie, all engagement whatever with Camilla Tyrold! I shall immediately acquaint my friends that henceforth . . . we Both are Free!" (641, original ellipses, original capitalization). But far from liberating them, this act seems only to further cement their bond. Edgar continues to lurk around, following Camilla and observing her at every turn. Camilla remains emotionally bound to him as well, thinking of little else, no matter where she goes or what painful social misadventures befall her.

The way that Camilla's character is described by the narrator suggests why the plot has the form and awesome length that it does. She is repeatedly contrasted with her cousin Indiana, who is uniformly described as either a two-dimensional or hollow three-dimensional object: "No statuary could have modelled her form with more exquisite symmetry; no painter have harmonised her complexion with greater brilliancy of colouring. But here ended the liberality of nature, which, in not sullying this fair workmanship by inclosing in it what was bad, contentedly left it vacant" (84). Camilla, in contrast, can only be illustrated by the effects that she has on others: "The beauty of Camilla . . . had an influence so peculiar on the beholder . . . it caught, by force and fire, the quick-kindling admiration of the lively; it possessed, by its magnetic pervasion, the witchery to create sympathy in the most serious" (84). It is true that the metaphors are somewhat mingled here, but the novel continues throughout to insist on representing her as an elusive, intimate substance that exists prior to and independently of her actions, speech, thoughts, or feelings. There is something inward that Edgar describes as "beaming in her eyes, playing in every feature, glowing . . . brightening" (234). This substance, however, cannot be directly brought into view. It can mostly be seen through the strong effects it has on other characters, both male and female, who swirl around her everywhere she goes with various degrees of admiration and intimacy.

Encouraged heavy-handedly by his tutor, Dr. Marchmont, Edgar follows Camilla from one place and situation to another, observing her behavior as she reacts to various people, situations, and propositions, and as others react to her. Her reactions are inconsistent, susceptible of various interpretations, and they are, as a rule, interpreted incorrectly or in the most damaging way possible. But the novel stresses the strange qualities of the stuff that is Camilla: "so unstable, as to be completely new moulded by every new accident, or new associate" (594). It is "inflammable," it is "pure, white, and spotless," but in a "dangerous state of first ductility"; it is "clear [and] transparent," or perhaps it is not (670). The authority figure, Marchmont, is largely to blame; he insists that "there is commonly so little stability, so little internal hold, in the female character" (654). Repeatedly, he advises Edgar: "By this test, then, prove your Camilla" (653). Marchmont concludes that "her character seems too unstable," citing its "wild vivacity . . . the changing propensities which now render her inconsistent to others, and fluctuating even to herself" (725). He blames the influence of the "contaminating world," which has seized her "before the character is fixed as well as formed" (726).

But this fluctuation is what Camilla is. Despite the novel's subtitle, "A Picture of Youth," it is not the case that she simply hasn't yet become herself. The quality that everyone finds so captivating is her pure ductility. At one point, Edgar ecstatically concludes, "She is still . . . the same; candid, open, flexible." But he then proceeds, "still, therefore, let me follow her. . . . I will watch by her unceasingly" (422). And he does. Although he wants Camilla to love him not for his external advantages but only for his interior substance, Edgar is aware of the essential unlovability of that substance: "chill, severe, repulsive" (653). Another character calls him "that frozen composition of premature wisdom" (375), which seems entirely accurate. Edgar continually melts, then freezes. His inner substance also seems incapable of holding a fixed form or direction. He keeps testing Camilla's internal composition to identify its fixed character, though he must know its essence is to be ductile, plastic. It is not too surprising when, at the height of her troubles, she expresses an intense desire to die. "A wish for death, immediate death . . . was the sole sensation. . . . To live, thus, seemed to her impossible" (839). Only in that way would she finally become the fixed, cool, unchanging substance whose qualities Edgar wants to confirm. Plot is the vehicle of this intense, intimate investigation. In its final paragraph, the novel distances itself from inductive reasoning, when the narrator blames Marchmont for having formed a "false . . . hypothesis" (913) about female character, based only on a few examples—a hypothesis which he then extended to Camilla. But Burney's plot itself performs a prolonged empirical investigation of substances at an intimate level.[39]

If the plots of Smollett's and Burney's novels are troubled by the instability or heterogeneity of material compositions, Radcliffe seems to embrace this ontology. One might wonder, though, about how to distinguish the plot of her novels from their other components, such as the texture of the descriptions that made *The Mysteries of Udolpho* so famous and make it still so striking. Radcliffe attentively traces gradations of change, particularly of color and light. *The Mysteries of Udolpho* seems particularly fascinated by the relative speed of change: how fast or slow, sudden or gradual, complete or incomplete it appears to be. Moreover, at every scale of time, the plot stresses repetition in a strange way. Its heroine Emily is always on the lookout for repetition, and yet she is always surprised when she finds something the same. I want to suggest that in *Udolpho*, processes of composition and decomposition—of change, mixing, and dissolving—make themselves apparent at surface levels of description and deep levels of plot. Such processes link together characters and setting, humans and inanimate substances, including heat, light, and sound. Instead of seeking a stable interior substance, or imagining a stable world in which action produces a change, Radcliffe's plot in this novel manifests a universe in which blending and dissolving, condensing and separating occur ceaselessly at an intimate, even molecular level. Change, then, is constant, and the deep plot traces such changes throughout the novel. In terms of the surface plot, actions of the usual human variety or scale (marrying, fighting, acquiring property, for example) come off as unwelcome distractions, ill-considered bustle, or interruptions. The plot trains its observation not on choice but on change, sometimes ecstatically and sometimes anxiously. The absence of change paradoxically takes on the nature of an event when it is discovered. Given the continual blending and dissolving of substances, the reappearance of the same thing (the same music, the same person, the same place) takes on the status of an almost miraculous resistance to these pervasive forces of change and mutual action.

A major focus of chemical investigation and debate in eighteenth-century chemistry was the nature of heat, which was related to the question of air.[40]

These were closely linked investigations. The theory of phlogiston held that things burned because they contained a substance called phlogiston, which was then released into the air. Those who denied the existence of phlogiston, like Antoine Lavoisier, had different ideas about what air was, and what happened in the process of combustion. In the latter's words, "We know that the property of burning is nothing else but the property which certain substances possess of decomposing vital air by the great affinity they have for the oxiginous [*sic*] principle."[41] But heat was thought to be a very subtle fluid that pervaded everything. Whether called phlogiston or caloric, heat was understood to be a kind of solvent. In Reill's words, "Matter was considered essentially fluid rather than solid": "Penetrability, solubility, and . . . internal combination overshadowed visions of impenetrability, collision, transferred motion, and the primacy of form."[42] Phase changes, how a substance could change from solid to liquid to vapor and back again, became a major topic. The translator's preface to Richard Kirwan's *Essay on Phlogiston* (1789) states, "The science of chemistry . . . informs us that this heat, or cause of temperature, whatever it may be, is the great opponent to the cohesive attraction, which, if not thus opposed, would prevent every change among the parts of bodies. When heat is communicated to bodies, they assume the various forms of expanded solidity, fluidity, and elasticity."[43] The world is seen here not as a miraculously balanced, harmonious system, but rather as a world that is frozen or locked into place. Heat and light are what account for change as well as sensation. Marcet explains that these are simply different ways of registering the same thing: "Light affects the sense of sight; Caloric that of feeling; the one produces Vision, the other the sensation of Heat."[44]

These natural philosophers can also provide striking descriptions of the qualities and tendencies of the material world. Lavoisier wrote in 1782, "It is necessary to consider all the bodies in nature as plunged in an elastic fluid of great rarity and lightness, known by the name of the igneous fluid, or principle of heat. This fluid, which penetrates all bodies, tends continually to separate their parts, and would separate them if they were not retained by their mutual attraction, that is to say, the attraction which is commonly called the affinity of aggregation."[45] Marcet faithfully conveys this picture: "One of the most remarkable properties of free caloric is its power of dilating bodies. This fluid is so extremely subtle, that it enters and pervades all bodies, whatever. . . . It dilates or expands a body so as to make it occupy a greater space than it did before."[46] The world appears dynamically constituted by a deep struggle: a "continual struggle between the attraction of aggregation, and the expansive power of caloric; and from the action of these two opposite forces, result all the various forms of matter."[47] Such processes could be observed through simple experiments, by noting changes in color, texture, and opacity. Marcet's teacher pours nitric acid on a piece of copper in a glass. "It has at present the appearance of a blue liquid," she remarks, "but when the union is completed, and the water with which the acid is diluted is evaporated, the compound will assume the form of regular crystals, of a fine blue colour, and perfectly transparent." Her student, Caroline, exclaims, "How very beautiful they are, in colour, form, and transparency!"[48] Joseph Priestley, in his investigations of air, reports that he discovered the existence of "dephlogisticated air" by observing how a candle burned in air that he had chemically produced: "What surprized me more than I can well express, was, that a candle burned in this air with a remarkably vigorous flame."[49] Qualitative observation of what is happening provides the key to what exists: "Though I did not give sufficient attention to the circumstance at that time, the flame of the candle, besides being larger, burned with more splendor and heat than in that species of nitrous air; and a piece of red-hot wood sparkled in it . . . and it consumed very fast."[50] In a different experiment: "A candle burned very strongly, and with a crackling noise."[51] Such changes lead him to conclude that "atmospherical air" is not a "simple elementary substance," but a compound.[52] It becomes critically important to notice how long such changes take to happen. This is Lavoisier's account of burning: an inflammable body "[decomposes] vital air . . . taking . . . the oxigene [*sic*] which was united to them. When this decomposition of the air is rapid . . . there is an appearance of flame, heat, and light; *when, on the contrary, the decomposition is very slow, and quietly made, the heat and light are scarcely perceptible.*"[53] *Udolpho* happens between these two extremes.

If, as the previous chapter argued, plot can serve as a unitary field of visibility, plot functions in *Udolpho* more like a shifting, enveloping atmosphere that solidifies and dissipates into various shapes and phenomena.[54] Landscapes, whether familiar or new, always end up dissolving: "Their outlines . . . melted from the eye into one rich harmonious tint" (7). "Through a distant window . . . were seen the purple clouds of evening and a landscape, whose features . . . no longer appeared distinctly, but, blended into one grand mass, stretched to the horizon, coloured only with a tint of solemn grey" (471). I would argue that these should not be dismissed as mere descriptions or subjective perceptions: they are presented as objective plot events *because* they do not occur within concatenated chains of causality. In *Udolpho,* matter is fluid, continuous, spreading apart, interacting continually with light and with sound. Light and sound are usually described in the process of slow, gradual mingling and dissolution: "The glow of the western horizon and the dark clouds, that came slowly from the east.... Above the darkness of the woods, her eye now caught a cluster of high towers, touched with the splendour of the setting rays, and, soon after ... she heard the faint swell of choral voices from a distance ... the strain had ceased ... arose from within ... rose by soft degrees ... swelled ... dropped into sudden silence ... was renewed ... died away, and was heard no more" (483). This cannot be considered a descriptive pause in the narratological sense, when the story or plot simply comes to a halt and nothing happens. Rather, such passages trace the gradations of an objective process of blending, composition and decomposition. Here, the components are light, color, and sound, but matter participates in the same processes; "The faded colours of [the pavilion's] painted walls and ceiling, and the decayed drapery of its once magnificent furniture ... abandoned to the empire of the changing seasons" (481) testify to the same slow, quiet decomposition. Similar phenomena, timed with greater precision, will appear in Austen.

We may not want to make too sharp a distinction, then, between atmosphere and action, as Jayne Lewis has suggested. Human action in Radcliffe seems to be an intensification, or a quickening of the forces that are present all around. Changes of light, color, heat, and sound *compose* the plot; they are not incidental to it, as we can see in a typical chapter (the twelfth chapter of volume 3). The Count de Villefort, Lady Blanche, and her lover St. Foix are traveling through the Pyrenees to La Vallée; the three characters are practically identical to St. Aubert, Emily, and Valancourt in the first part of the novel, and the journey is also nearly identical, only made in the reverse direction. The same melancholy and merry sounds are heard, and magnificent descriptions of changing light are given repeatedly: "The rays of the setting sun now threw a yellow gleam ... soon, even this light faded fast ... a dark mass of mountains now alone appeared.... A melancholy gleam still lingered on the summits" (597). The repeated "now" reminds us that something is happening; this is not static description, happening in a suspended moment, but an event with an objective duration. Blanche alternates between apprehension and "pensive pleasure, as she watched the progress of twilight gradually spreading its tints" (598). The travelers stop; a fire is made, and St. Foix observes the mingled effects of flame and lightning. There is the "momentary light" of the latter, along with the fire's "partial gleam": "It threw a strong, though pale gleam, and glittered on [their guides'] bright arms; while upon the foliage of a gigantic larch . . . appeared a red, dusky tint, deepening almost imperceptibly into the blackness of night" (601). The travelers are lost; they become part of a classic Radcliffean assemblage: a dog barking, ruined towers, broken paths, voices, a watch-tower. They are admitted into the tower by people who turn out to be smugglers; Blanche and her party are in peril; they are rescued (again) by the sudden, reliably unexpected appearance of Ludovico. Interestingly, the smugglers, whom Blanche overhears discussing whether to murder them, are not sure of St. Foix's identity (one thinks he recognizes him). They are chiefly motivated by the glittering of his ring and of the diamonds on the miniature that Blanche wears (613). The sequence of actions inside the watchtower, though involving people and plans, rescue and escape, does not materially differ: noises interrupt, light partially reveals, and things appear and disappear, slowly or suddenly, reflecting or absorbing the light. Earlier in this chapter, the Count had entertained his companions by telling them "the natural history of the scene, among which they wandered. He spoke of the mineral and fossile [*sic*] substances . . . the veins of marble and granite . . . the strata of shells, discovered near their summits" (602). He then seamlessly moves on to "events and circumstances, connected with the civil story of the Pyrenées . . . celebrated sieges and encounters in early times" (602). Human-scale events are blended into the landscape, differing mostly in terms of their speed or concentration; their effects have also faded away. "While she considered, that she was on the very ground, once polluted by these events, her reverie was suddenly interrupted by a sound, that came in the wind" (602).

Given the constant dissolution of the world, "very slow and quietly made," in Lavoisier's words, it becomes easier to understand the novel's attitude toward the repetition of phenomena. Emily always seeks it, yet is shocked to find it. "Emily, as the sounds drew nearer, knew them to be the same she had formerly heard at the time of her father's death . . . she was so much affected, that she had nearly fainted" (525). Returning to Toulouse, she is overcome to see "the very avenue," the same house, "the same high trees . . . the same flowery thickets . . . the very plants" (583). The greatest shock, though a welcome one, is the discovery that Valancourt, after his long absence and despite his own confession to Emily that "I am not the same!" (513), is in fact unchanged, the same. In the final chapter, as Emily and Blanche are both married "on the same day . . . old Dorothée . . . sighed, and said, the castle looked as it was wont to do in the time of her youth" (671). Sameness is not the absence of an event. It is a surprising event, one that beats the odds. It is the unlikely re-creation or reenactment of a particular configuration in a fluid world, a process of decomposition or analysis run in reverse.

Changing, blending, and dissolving provide the deep plot of this novel. It is true that the novel does not give up the idea of the type, but here the type is equated with limitation, flatness, and stasis. It thus comes to be assigned to minor characters such as servants (indeed, type becomes the essence of minorness), while protagonists escape that kind of determination. It's also true that the plot does not neglect causal explanation of the more usual kind, but the redundancy of such explanations was obvious to Radcliffe's readers. Does it really matter how or why Ludovico suddenly disappears and appears when so many things have the same habit of doing so, only more gradually? The famous object behind the black veil, however, merits a closer look: "a human figure . . . partly decayed and disfigured by worms" (662). If Emily had looked again, "She would have perceived, that the figure before her was not human, but formed of wax" (662). What is most unnatural about this figure is that it freezes the process of decomposition going on all around it. In doing so, it is both like and unlike Emily herself: "As she listened, she was chilled with superstitious awe . . . Emily, turning her eyes from the massy darkness of the woods . . . saw, on the left, that effulgent planet . . . she unclosed the casement to listen to the strains, that soon gradually sunk to a greater distance. . . . Chilled with a melancholy awe, she retired once more to her bed" (84). The absence of heat, the presence of light, and moving sounds are all part of this plot's intimate composition.

CHAPTER 6

Molecular Possibility in Austen's Plots

Miss Bates, in Austen's *Emma*, appears to be a character without much claim to plot. She is an older single woman, neither handsome, clever, nor rich; there is no marriage in her past or future. She only talks. Her sentences famously have no beginning or end, head or tail; her chatter seems to consist of disjointed trivia. But there's something about her talking that deserves our attention. While there are plenty of silly or tedious characters in Austen's novels, we tend to hear from their lips only a sample of their speech. The narrator summarizes for us, as when Mr. Collins accompanies the Bennet sisters on their walk: "In pompous nothings on his side, and civil assents on that of his cousins, their time passed till they entered Meryton."[1] The speech of Miss Bates, however, is quoted directly, and for remarkably long stretches. Why? There is a quality to her speech that defies summary: it has to be exhibited. And, in turn, it reveals something fascinating. In the second volume of the novel, Emma and Harriet visit a store in Highbury. "Voices [approach] the shop—or rather one voice and two ladies"; from that point Miss Bates's monologue continues until the end of the chapter, occupying close to half its length. Here is an extract: "Patty came to say she thought the kitchen chimney wanted sweeping. Oh! said I, Patty do not come with your bad news to me. Here is the rivet of your mistress's spectacles out. Then the baked apples came home . . . what is our consumption of bread, you know? . . . [Jane] really eats nothing—makes such a shocking breakfast, you would be quite frightened if you saw it. I dare not let my mother know how little she eats—so I say one thing and then I say another, and it passes off . . . I have so often heard Mr. Woodhouse recommend a baked apple."[2] Miss Bates's speech may strike us as familiar, the essence of the everyday. Austen's own letters, William Galperin has shown, bear a resemblance to this character's way of talking.[3] Yet—perhaps because of this very quality—her speech cannot be gathered up into a phrase like "pompous nothings." It does not rely on social roles or respect social boundaries. Even though Miss Bates frequently acknowledges her listener (you can practically feel her grabbing your arm), her speech overflows. It exhibits a field of minute, chaotic, incessant movements in which rivets, apples, and spectacles, the actions of eating and recommending bounce around, collide without losing any of their motion, and come back again, swirling around endlessly.

Twelve years after *Emma's* publication, a Scottish botanist observed a strange motion under his microscope as he viewed some pollen grains in water. The particles were constantly moving around as well as bending and stretching. Extending his investigation to other plants and then to inorganic matter, including minerals, dust, and a fragment of the Sphinx, Robert Brown concluded that "these molecules . . . were always found to exist" in everything.[4] As the physicist Jean Perrin described it in 1910, Brown saw that "all the particles situated in the liquid instead of assuming a regular movement of fall or ascent, according to their density, are . . . animated with a perfectly irregular movement. They go and come, stop, start again, *mount*, descend, *remount again*, without in the least tending toward immobility. This is the *Brownian movement*."[5] It took until the end of Brown's century for physicists to realize "in this, supposed insignificant, phenomenon, a fundamental property of matter . . . molecular movements were able to give the explanation of the phenomenon."[6] Perrin makes several points about Brown's discovery that are potentially relevant to the way that Austen conceives of plot. First, a mass that may "appear completely motionless" is actually in constant internal motion. Second, "the particles in suspension are agitated the more briskly the smaller they are." And finally, "What is really strange and *new* in the Brownian movement is, precisely, that it never stops. At first that seems in contradiction to our every-day experience of friction." And yet this is the case: equilibrium is only "a statistical equilibrium. In reality the whole fluid is agitated indefinitely and *spontaneously* by motions the more violent and rapid the smaller the portion taken into account."[7]

Perrin's description can serve as an account not only of Miss Bates's speech, which never stops, but of Austen's plots. (Despite the anachronism here, it is perhaps not too far-fetched to draw this comparison if we recall that the Austen household did possess a microscope, and that Austen's novels were often described in such terms.[8]) It is the more apt because equilibrium is a metaphor often used by theorists of plot.[9] Here, though, it reveals something important to the novels I have been examining: that equilibrium is assumed rather than real. The agitations of plot are spontaneous, ongoing, and potentially observable even in the smallest portions of narrative. We see the grains, the molecules of plot swirling around in the field of Miss Bates's speech. The plots of Austen's novels turn on interactions between bodies of matter conceived on a range of scales, down to the nearly microscopic.[10] The human is only one of these scales. As in Miss Bates's speech, heterogeneous entities collide with and displace each other almost without cease. They do not even need to be solid things: a space may open up on a bench, or a sound may result from a pen that drops to the floor, both of which happen in *Persuasion*. But as a result of tiny movements like this, large things happen. Bodies can break apart and reaggregate, for a time. We are so absorbed in watching this happen that we may not seek out the causes. Indeed, the causes seem to be contained within the movements of these bodies. As Perrin notes, "There is . . . an agitation maintained indefinitely without external cause."[11]

I hope to make a somewhat different point from what many have noticed: the attention to small, everyday things that made her earliest critics compare Austen's novels to "Flemish paintings."[12] My point is that each of these tiny things represents something like a unit of plot, if we can thus adapt Peter Brooks's model of narrative.[13] Not only Miss Bates but even the major characters in Austen's novels care deeply about such small things; they are infused with desire. But even more than the things themselves, their actual and possible movements are the focus of the keenest concern. It's easy to laugh at the pencil-stub that Harriet treasures because it came from Mr. Elton's pocket, or Lady Catherine's views on where Charlotte's furniture should go, but even serious characters like Mr. Knightley and his brother can absorb themselves in questions such as "moving the path to Langham, of turning it more to the right" (85).[14] Or who will go in the carriage with whom? This has to do with more than questions of property, of who will own something, and it also goes beyond the sentimental associations that objects may possess. It is linked, I believe, to the way that these fictional worlds are materially composed, the sense of constant small-scale agitation that cannot be stopped, of collisions, cohesions, and displacements. While many readers, even very sympathetic ones, seem to find Austen's world impoverished in terms of plot, I see instead a striking fullness, even a crowdedness. There is plot in every inch of it. It is not only Anne Elliot who experiences a room as "full—full of persons and voices."[15] From Catherine Morland's first ball in Austen's first novel to Fanny Price's home in Portsmouth, everywhere we see, in Emma's phrase, "a crowd in a little room . . . dancing without room to turn in" (195). This teeming, jostling quality seems to be an aspect of the plot's physical world as well as the social one. Recall Perrin's description of Brownian movement: "They go and come, stop, start again, *mount*, descend, *remount again*, without in the least tending toward immobility." It means that when one thing moves, it will probably bump up against something else. Even the smallest displacement is an event with consequences. Right after the passage quoted above, when Miss Bates is interrupted, "Emma wondered on what, of all the medley, she would fix" (186). Any piece of it can be a story.

The plots of Austen's novels invite a detached view.[16] This is not an emotional detachment: feelings of empathy, comfort, and even love were and remain available to her readers, as Deidre Lynch has shown.[17] Rather, it is more of a cognitive one: an invitation to view things, bodies, and their movements abstractly—we could call it a natural philosophical perspective. While characters are highly individuated, they are also subject to forces that make them appear, to borrow a phrase from one of Newton's popularizers, as "body in general."[18] The plots consistently foreground the question of change as such, applying it to widely different kinds and scales of entities, both human and nonhuman (furniture, trees, Darcy). They trace change as it happens at speeds ranging from a flash of knowledge or insight ("the speed of an arrow"), to the time it takes to reread a letter. Or it may require seven years of reflection in order to "[think] very differently."[19] While change arguably lies at the heart of any narrative, Austen's novels ground it in the small movements and collisions of material bodies, rooting it deeply into the implied composition of the represented world.[20] Even Fanny Price, the most static of protagonists, has this insight as she's about to be sent back to Portsmouth: "She began to feel that she had not yet gone through all the changes of opinion and sentiment, which the progress of time and variation of circumstances occasion in this world of changes."[21] She foresees, in an odd moment, a future self with which she will not coincide. Nothing is exempt, no matter how small or how important; change thus acquires a kind of priority as an abstraction on which the plot's coherence depends. But the pervasiveness of change relies directly on the fullness, the dynamic crowdedness of the physical substrate. Any collision, addition, removal has consequences on neighboring particles. And it is perhaps this inexhaustible quality that has allowed Austen's novels to serve, in Lynch's words, as "source material" for so many projects of both individual and national desire.[22]

Austen's relationship with all the branches of natural philosophy was broad, comfortable and well-informed, as Peter Knox-Shaw has shown; its instruments and practices were part of her world and were taken up around her with enthusiasm. Her brother, James Austen, for example, studied and wrote about chemistry and natural history; the periodical that he published for a time discussed developments in chemistry and mentioned Lavoisier, Priestley, and Kirwan, among others.[23] Her novels were written in a period, Jan Golinski argues, when science (chemistry in particular) continued to be publicly available for middle class audiences, including women. Natural philosophy retained a commitment to accessibility, and rested its claim to legitimacy on its continuity with other arts, pursuits, and concerns, rather than setting itself apart from them.[24] Austen's novels' obvious delight in surface variation also led contemporary readers to make the connection with natural history; Richard Whately, for example, praises the delineation of her minor characters: "To the eye of a skilful naturalist," he remarks, "the insects on a leaf present as wide differences as exist between the elephant and the lion."[25] Knox-Shaw points out that Austen's novels were particularly popular with men of science, including William Whewell.[26]

But this chapter is not about Austen's direct relationship to science. Instead, I want to suggest that by this point in the history of the novel, when Austen can name and defend the genre as such, the views, assumptions, and practices of natural philosophy in its multiple branches have created a kind of ground floor for the way that novels represent worlds. As we saw in the previous chapter, Burney and Radcliffe seem to build their plots on the way that complex substances, including light and heat, seem to combine with and alter one another. Austen explicitly cites Burney's *Camilla* and Radcliffe's *Mysteries of Udolpho*, as well as Edgeworth's *Belinda* (discussed in the next chapter) in a famous moment of *Northanger Abbey*, describing them as works "in which the greatest powers of the mind are displayed, in which the most thorough knowledge . . . [is] conveyed to the world."[27] What Austen seems to acquire from these novelists, then, is this sense of a world as an intimate material composition, as well as a way of viewing its changes as eventful. I want to suggest that the plots of Austen's novels rely deeply on such small, material movements, an unceasing agitation that sweeps through and changes all configurations. Even when we see the symmetry of a novel's plan, as in the two volumes of *Persuasion*, what allows the plot to accomplish itself is this molecular movement. The larger events depend on these small collisions, jostlings, displacements, dilutions. As Emma believes she herself has done, the plot has to manage these "many little matters," or else "it might not have come to any thing at all" (11). Note that the event of marriage, often used as a metonymy for plot itself, is literally a reaggregation of bodies taken from one place and brought together in another. (*Emma* begins by emphasizing that fact; Miss Taylor's marriage is seen as her removal from Hartfield.) But marriage is only one among more prominent events. Austen's plots, like Defoe's, seem to take a particular pleasure in the smallness, multiplicity, and heterogeneity of intermediate causes.

Classical models of plot highlight proairesis, rational decision, and action. We find an example of such a plot in concentrated form in the story of Mr. Weston, told at the beginning of *Emma*. That story comprises his interest in Miss Taylor, his goal of marriage, and his intention to purchase Randalls: "He had gone steadily on, with these objects in view, till they were accomplished. He had made his fortune, bought his house, and obtained his wife" (14). Small plots such as this one, family or individual histories, can be found embedded in the novels, often to set the stage for a character's appearance. But a story like this does not much resemble how Austen's plots are formed. Much of their interest and attention goes to what Defoe called "the treasury of immediate cause."[28] Take the example of Harriet and Frank Churchill, in the third volume of *Emma*: "A fine young man and a lovely young woman thrown together in such a way, could hardly fail of suggesting certain ideas to the coldest heart and the steadiest brain . . . could even a mathematician have seen what she did . . . without feeling that circumstances had been at work to make them peculiarly interesting to each other?" (263). The circumstances are many, intricate, and diverse. They include the "sudden turn" made by the Richmond road, "a stout woman and a great boy," "a steep bank," the cramp in Harriet's leg, brought on by too much dancing the previous evening, Frank Churchill's chance delay in leaving Highbury, and a pair of scissors that he had borrowed from Miss Bates that he had "forgotten to restore" (261). This is the kind of plot that seizes the imagination of all Highbury, "the young and the low . . . all the youth and servants" especially. It is the story that Emma's nephews ask her to tell "every day . . . tenaciously setting her right if she varied in the slightest particular from the original recital" (264). Each small cause, including the cramp and the scissors, is essential to this story of dramatic rescue. A more serious example is found in the visit to Sotherton in *Mansfield Park*: a key left in the house, and a small gap at the edge of a locked iron gate together release Henry Crawford and Maria into the park so "that their views and their plans might be more comprehensive" (91); the consequence is their elopement at the novel's end. A last example of a minute but consequential cause can be found in the walk at Lyme Regis in *Persuasion*, when a momentary miscalculation of "half a second"—the more ironic because the characters had just decided, "By all their calculations there was just time for this [last walk]" (91)—leads to Louisa's fall, her prolonged illness and her subsequent change of personality, which frees Wentworth to return to Anne.

Social and psychological phenomena are obviously crucial to Austen's novels: what happens between people, or inside their minds, as the following chapter will discuss. But people are also viewed in another way. There seems to be a level at which bodies interact as natural phenomena—that is to say, as equally material, though not identical to each other. This is Boyle's view, as described here by Steven Shapin: "Matter and motion were . . . simple and finite in themselves but capable in combination of producing almost endless diversity. . . . All the experienced diversity of natural objects was thus to be accounted for by the mechanically simple and primitive qualities that necessarily belonged to all bodies as bodies, and not to roses or iron bars or magnets as types of bodies."[29] We can think of plot in Austen's novels as that level, where the movement of one body results in another's motion. They move and are moved; individual characters exist within aggregates and as themselves material aggregations that are open to rearrangement. Things happen because of the situations and motions of these bodies and the ways in which they cannot help interacting with each other. Certainly, other aspects of the narratives perform a ranking or ordering: for instance, the mechanisms of narrative attention that ensue in what Alex Woloch has called "character-space."[30] The inner thoughts and feelings of some characters are explored in far more detail than those of others. But when it comes to what happens in the plot, and to the tracing of changes, movements, and combinations, both virtual and actual, a greater equality seems to pertain. The mimetic work of plot is not limited to social phenomena such as class distinction or hierarchy. Austen's plots also imitate a world of matter that is structured by certain rules governing movement, space, time, and number. "All of these motions [of natural phenomena]" we recall, "are performed by certain rules, and always subject to the same laws . . . even in those things which appear to us wholly fortuitous and uncertain, certain rules are without doubt observed."[31] There remains a sense that there are inescapable, objective regularities in the way that things happen. These can be thought of as the material medium through which consequences are transmitted.

A deeper, less human plot transcribes itself through such phenomena, as well as an idea of objectivity not reliant only on style or discourse.[32] Austen's plots contribute to an idea of objectivity as a material nexus of extremely small, nearly invisible causes that add up. This brings us to the threshold of what Ian Hacking has described as "the underlying metaphysics of probability" in nineteenth-century statistics: "the picture of minute causes" that result in large-scale regularities that govern human populations.[33] But Austen's novels give us these in narrative form, describing not the large-scale equilibrium but the tiny fluctuations, the constant agitations.[34]

As we saw in the novels of Radcliffe and Burney, Austen's plots focus on describing change more than recounting actions. But the bodies or entities in her novels possess boundaries that are more definite, clearer than in Radcliffe, though varied and always subject to alteration. Bodies are configurations with a certain "shape, size, arrangement, and motion of the material constituents."[35] But—and this is key to the way her plots work—they are temporary aggregates. While the boundaries of individuals seem clear, they too are open. Knox-Shaw observes that "porousness is the key to Jane Austen's sense of self-making."[36] Austen noted that "seven years . . . are enough to change every pore of one's skin, and every feeling of one's mind."[37] That account, however, suggests unconscious transformation. But Austen's plots assert the clarity and solidity of boundaries in order to undo them. A change in the boundary or composition of an aggregate is an event that leads to other changes. It can be as small as an impulse crossing someone's mind, or can involve a visible change in position—moving to a different spot on a bench, leaving a few inches on the end (as Anne Elliot does). Obviously, arrivals and displacements are events, but so are gaps or vacancies. The vacancy at Netherfield is arguably as important an event in *Pride and Prejudice* as the arrival of the militia.

An event can be a virtual change as well as an actual one; boundaries are temporal and modal as well as spatial. Virtual events, thought of as necessary, possible, or desirable, reveal where the boundaries of bodies happen to be situated at that moment. At one point in *Mansfield Park*, Fanny Price thinks she's going to be sent to live with Mrs. Norris. Even though it doesn't happen, it reveals that she has already at this point become part of the Bertram body, perhaps indeed the very subtle, invisible, yet resistant thing that holds together that self-divided, disjointed, unruly body. Plot shuffles bodies and imagines various combinations: for example, Elizabeth with Collins, Emma with Elton, Fanny with Henry. There is Anne Elliot within the Musgrove body—a system that comprises rhythms of movement and measures of space, as well as several connected households. There is also what we might call the greater Highbury organism, with its strange, teeming mixtures. The arrival of Frank Churchill, when it fails to happen, shows the degree to which Emma unconsciously regards herself as part of that aggregate body.

For the most part, Austen treats bodies, large and small, as bounded material accretions whose parts are in constant motion. The boundaries between such bodies are where plot can be said to occur. On one level or scale, we see phenomena such as "collision, transferred motion, and the primacy of form."[38] They can transmit motion from one to another in a concatenated sequence. In *Pride and Prejudice*, the large-scale sequences of the plot depend on a conception of bodies as receiving and transmitting force. They rely on blind attraction and blind action: action that is almost like a blow or a strike. From a psychological point of view, the plot is tautological. Jane and Bingley love each other because they love each other. More importantly, there is force, and resistance to force. The initial force lies in the character of Darcy, who seems to be the unmoved mover: "He walked here and he walked there, fancying himself so very great!" (9). There is little more to this character than "the power of doing what he likes" (140), situated in a material body. In Newton's definition of power, "A power is any force acting upon a body to move it."[39] For Darcy, this is the power to direct not only his own motions but also the movements of Bingley—and of Wickham, Lydia, his sister, housekeeper, tenants, and so on. Elizabeth remarks to Colonel Fitzwilliam, "I imagine your cousin brought you down with him chiefly for the sake of having somebody at his disposal" (141). There is a discernible bitterness beneath the playful observation, even before she learns of his involvement in the termination of Bingley's courtship. "I do not see what right Mr. Darcy had . . . to determine and direct in what manner [Bingley] was to be happy" (143). The scandal is not really a question of what moral right he had, but that such force or power should exist in a human world, that anyone should possess such "boundless influence" (143). Darcy is otherwise not very animate; he possesses minimal interiority, like many Austen characters: "Why Mr. Darcy came so often to the Parsonage, it was more difficult to understand. . . . He seldom appeared really animated." All he seems to show, paradoxically, is "nothing but absence of mind" (139). His conversations with her at this point consist of "an awkward pause" followed by "odd unconnected questions" (140). Until the third volume, Darcy is little more than a material body possessed of force that is sufficient to "separate Mr. Bingley and Jane . . . he was the cause . . . the cause of all that Jane had suffered, and still continued to suffer . . . no one could say how lasting an evil he might have inflicted" (143). Elizabeth imagines the effects of that violent separation continuing indefinitely.

The attraction of characters—of Darcy to Elizabeth, Bingley to Jane—is represented as "a mechanical consideration of bodies."[40] It is a change in "the motion of material constituents," a potentially altered arrangement of constituents in the larger Bennett-Bingley-Darcy collective body.[41] Elizabeth "attracted [Darcy] more than he liked" (44), so she is quite correct when she tells him, "You liked me against your will, against your reason, and even against your character" (146). The attraction of Bingley to Jane is equally innocent of interiority: "The beauty of her sister re-kindled the admiration of her former lover. When first he came in, he had spoken to her but little; but every five minutes seemed to be giving her more of his attention" (257). It is the attraction of surface to surface. Once Darcy's force is removed, Bingley keeps rolling onward to his natural end: "Elizabeth ... now smiled at the rapidity and ease with which an affair was finally settled, that had given them so many previous months of suspense and vexation" (265). Bingley and Jane are both opaque or transparent material bodies that can be moved, stopped, or held back in the course of gravitating toward each other: "When we use the Words Gravity, Gravitation, or Attraction; we have a Regard not to the Cause, but to the Effect; namely to that Force, which Bodies have when they are carried towards each other."[42]

Darcy and Elizabeth alternate between being active channels of force and passive recipients of it. The moments of impact are precisely delineated. Elizabeth rebuffs him, pushes him away and into a different line of motion. Like a physical blow in Fielding, it is the work of one second, though the after-effects are lasting: "Your reproof, so well applied, I shall never forget: 'had you behaved in a more gentleman-like manner.' Those were your words. You know not, you can scarcely conceive, how they have tortured me" (281). She responds, "I was certainly very far from expecting them to make so strong an impression" (281). But that push and nothing else is shown to be the efficient cause of the startling change revealed at Pemberley: "The change was so great, and struck so forcibly on her mind, that she could hardly restrain her astonishment from being visible" (199). The metaphors of physical force do not seem entirely metaphorical, as the reference to her outward comportment suggests. At that point, she's still not sure whether this change in Darcy's composition, this new internal arrangement that results in "complaisance" rather than "self-consequence" (199), is temporary or permanent. She feels skepticism: "My reproofs at Hunsford could not work such a change as this" (193). On this point, she happens to be wrong. But then Elizabeth is also a different arrangement herself, as the result of Darcy's letter. He asks her about it later, questioning the speed of its effect: "Did it *soon* make you think better of me? . . . She explained what its effect on her had been, and how gradually all her former prejudices had been removed" (281, original emphasis). When he asks her to burn the letter, she remarks, "The feelings of the person who wrote it, and the person who received it, are now so widely different from what they were then" (282). Not only the feelings, but the persons themselves are different, as Darcy notes: "Such I was, from eight to eight and twenty; and such I might still have been but for you . . . !" (282). Elizabeth's blow—her "reproofs at Hunsford"—turns him not only into a new direction but practically into a different body. Inertial movement is a powerful thing: "We see that bodies by their nature are inactive and incapable of moving themselves; wherefore unless they be moved by some extrinsical agent, they must necessarily remain for ever at rest. A body also being once in motion, continues in motion according to the same direction, in the same right line, and with the same velocity, as we see by daily experience; for we never see any change made in motion, but from some cause."[43] As in a Fielding brawl, punch and counterpunch follow each other, changing the trajectory of each body. Elizabeth hits first; Darcy responds with the letter; both stagger, rearrange themselves, follow a different course. The plot also gives due weight to smaller, intermediate causes: notably, the Gardiners, "the persons who, by bringing her into Derbyshire, had been the means of uniting them" (298), and Lady Catherine, whose dramatic intervention has an "effect . . . exactly contrariwise" to her intention to separate them (280). The enchainment of these movements creates a pleasing effect similar to what Desaguliers describes: "It appears by Observation that Gravity is the Cause of the Fall of heavy Bodies, which observe certain Laws in their Motion . . . a heavy Body by its Descent moves the Axis of a Wheel, that carries round another by its Teeth, which by the Intermediation of other Wheels and Pinions, carries round a Hand upon a Dial-Plate to measure Time, or for other Uses."[44]

But *Pride and Prejudice* stands somewhat apart from the other novels in its suggestion of the passivity of matter. In *Mansfield Park* and *Emma*, by contrast, we see subtle matter in constant motion at a very small scale. Characters are comprised of ratios of matter and motion. The apothecary Mr. Perry moves so fast and contains so little matter that he is never directly seen. Mr. Weston is "a man who had been in motion since eight o'clock in the morning" (237)—appropriate for the father of Frank Churchill. Frank exists for much of the novel as a vector of movement, an arrow pointing away from Highbury or toward Jane. He comes tomorrow, he is early, he is delayed. His exact location is often unknown. When others speak of him, they mention "the route and the method of his journey," his speed and direction (148). Even his father notes that "the sort of constant expectation there will be of his coming in to-day or to-morrow, and at any hour, may be more friendly to happiness than having him actually in the house" (242). Frank is more virtual than present, an "eagerness to arrive which had made him alter his plan, and travel earlier, later, and quicker" (149).

Even more illustrative are the three Ward sisters in *Mansfield Park*. Lady Bertram, "a woman of very tranquil feelings, and a temper remarkably easy and indolent" (6), is glued by inertia to her sofa. She never leaves it voluntarily. "She was a woman who spent her days in sitting nicely dressed on a sofa" (20). She does not educate her daughters; she "did not go into public with her daughters. She was too indolent even to accept a mother's gratification in witnessing their success" (34). In contrast, Mrs. Norris never ceases moving, "walking, talking, and contriving" (9), not only setting things in motion but hurrying them along with the transfer of all her own momentum: "I will write to my sister to-morrow, and make the proposal . . . I will engage to get the child to Mansfield" (9). She contrives Maria's disastrous marriage to Mr. Rushworth. "Mrs. Norris was most zealous in promoting the match, by every suggestion and contrivance . . . among other means, by seeking an intimacy with the gentleman's mother . . . and to whom she even forced Lady Bertram to go through ten miles of indifferent road, to pay a morning visit" (37). Attending these two bodies, Fanny "either sat at home the whole day with one aunt, or walked beyond her strength at the instigation of the other; Lady Bertram holding exercise to be as unnecessary for every body as it was unpleasant to herself; and Mrs. Norris, who was walking all day, thinking every body ought to walk as much" (35). The third sister, Fanny's mother, appears to be the exact average of the other two: "Her days were spent in a kind of slow bustle; always busy without getting on, always behindhand and lamenting it, without altering her ways" (361). By nature "easy and indolent, like Lady Bertram's," Mrs. Price is unhappily placed in a situation requiring "exertions" and motions that she is not capable of performing (361). It is revealing that Austen makes this third sister not a virtuous or golden mean, but an unhappy one. It is in keeping with her determined anti-idealism, her refusal to moralize, and her view of the world as containing no exceptions. Mrs. Price is disappointing to Fanny, but the conduct or behavior of all three of these women arises from "the motion of the material constituents," in Shapin's words (46). Those material constituents also include the world around them: calm, spacious, and ordered in one case, swirling, loud, and chaotic in the other.

The place or position that something occupies is significant in Austen's plots, but not because it indexes an ideal social order. Rather, it matters what position within an aggregate (temporary or more lasting) you occupy because of the plot possibilities it holds: the collisions, conjunctures that could ensue. Indeed, a gap opening up is an important event; there are few things more eventful in Austen's plots than a vacancy of any kind. On the drive to Sotherton, "The place of all places, the envied seat, the post of honour, was unappropriated. To whose happy lot was it to fall?" (75). Powerful affects belong to the positions rather than to the people: to be within is to be in gloom, to sit on the box is exhilaration. Fanny takes the place of the Miss Bertrams after Maria's marriage, "Becoming as she then did, the only young woman in the drawing-room, the only occupier of that interesting division of a family in which she had hitherto held so humble a third" (190). Fanny also takes the place of Maria Bertram in Henry's affections, for a while, at least. And of course, Susan, Fanny's sister, "remained to supply her place" after Fanny's marriage: "First as a comfort to Fanny, then as an auxiliary, and last as her substitute, she was established at Mansfield" (438).

Such shifts in place or position are not subordinate to more important events but are themselves the event. Many scenes of group-walking seem to make this point. We watch a collective body continually shift its composition, moving in varied rhythms, paces, and combinations. And the configuration matters as much as anything else in the scene. Walking with Wentworth, Louisa learns the pleasure of jumping down (as well as of being unpersuadable), which results in her fateful leap during their last walk at Lyme. Anne is reminded of the importance of overhearing; marginality has its uses. There are also more subtle implications for larger collective bodies: for instance, during the long sequence in the wilderness at Sotherton, Fanny largely sits still on the bench, forgotten by all the walkers, but she acts as the mechanical center of motion and the center of gravity: "In a mechanical consideration of bodies," Benjamin Martin writes, the center of motion is "that point . . . in a solid body that remains at rest, while all the other parts of the body move about it," while the center of gravity is "that point which being supported, the whole body is supported, or kept from falling. The manner of finding this point in any body, is best shewn by experiment."[45] The Sotherton walk can be seen as an experiment through which the plot reveals that Fanny is indeed the center of motion, that which keeps the whole body of Mansfield from falling, though no one knows it yet.

The walks seem to show in concentrated form the essence of Austen's plots. They are choreographed to suggest an element of randomness in the shifting and movement of bodies, the rearrangement of boundaries. At Sotherton, Donwell Abbey, and Box Hill, there are clear social and psychological motives that underlie the characters' motions. But something also happens at a level that eludes anyone's understanding and even defeats their intentions. Henry Crawford summarizes Sotherton thus: "It was a hot day, and we were all walking after each other and bewildered" (226). The swarm of minute material causes takes over, even guiding the movements of the characters. Box Hill is the most striking example. Despite the "very fine day" and "all the other outward circumstances of arrangement" that were well handled,

> There was deficiency. There was a languor, a want of spirits, a want of union, which could not be got over. They separated too much into parties. The Eltons walked together; Mr. Knightley took charge of Miss Bates and Jane; and Emma and Harriet belonged to Frank Churchill. And Mr. Weston tried, in vain, to make them harmonize better. It seemed at first an accidental division, but it never materially varied. Mr. and Mrs. Elton, indeed, showed no unwillingness to mix . . . but during the two whole hours that were spent on the hill, there seemed a principle of separation between the other parties, too strong for any fine prospect, or any cold collation, or any cheerful Mr. Weston, to remove. (288–89)

There seemed a principle of separation: this nearly mechanical way of putting it suggests the existence of something apart from the intentions and exertions of any characters, including Emma, some kind of quasi-natural law of motion. The mixture keeps separating because of "circumstances of arrangement" that are too small, too intimate to be visible to the eye.

The final walk at Lyme in *Persuasion* offers another example in which something goes wrong in the composition of the aggregate, but the cause cannot be pinned down to one object within it: "Loitering about a little longer, they returned to the inn . . . *something occasioned an almost general change* amongst their party, and instead of Captain Benwick, she had Captain Harville by her side . . . as they drew near the Cobb, there was such a general wish to walk along it once more, all were so inclined . . . that the difference of a quarter of an hour, it was found, would be no difference at all" (85–91, emphasis added). Insofar as there is a cause, it seems to be an empty space within the configuration, a "something" that causes "an almost general change," which I understand to mean that the whole group shuffles. Yet this internal spatial rearrangement of components seems spontaneous, its own cause, a kind of momentary "general wish" that takes over everyone. They are all attracted to each other in a new way. This walk, of course, ends with Louisa's mistimed jump, which will rearrange the couples once more and permanently.

Austen's plots are also full of matter that gets in the way of what characters want: Mr. Elton starts talking to Emma just when she wants to overhear a conversation about Frank Churchill on the other side of the table. Frank Churchill tells Emma to watch him while he speaks to Jane Fairfax, and then places himself exactly so as to block her vision. Fanny Price refuses inexplicably to go along with plans that everyone else wants to pursue, whether acting or marriage. Everywhere, someone or something is demanding, distracting, or blocking what seems to be the important thing, the thing you are intent on seeing or understanding. But this view is itself something of a trap, most obviously in *Emma*. The plot is not what is happening behind the noise or the screen: it is itself the noise, the friction, the distraction, and the delayed or mediated transmission.[46]

Causal effects can be muted, delayed, buried, or relayed along a network of smaller causes. That one step forward and closer to another body, or the few inches of space made somewhere, the sound that gets in the way of someone trying to hear something: such phenomena are more than just the setting, occasion, environment, or backdrop for intentional action or decision. The distinction here between event-accidents and the subtle matter that seems to fill all the spaces in Austen's plots may come down to a difference of scale. Something transmits effects: Louisa's contact with the ground, "the fall from the Cobb, might influence her health, her nerves . . . her character to the end of her life" (136). When someone slams a door, Louisa Musgrove "starts and wriggles" (176). The "fine wind blowing" restores Anne, reanimates her; matter transmits vibrations from the external world to the surface and the inside of the body: "The almost ceaseless slam of the door, and ceaseless buzz of persons walking through" causes Anne "to breathe very quick, and feel an hundred things in a moment" (148). Her most intense feeling is described as "senseless joy" (136), a kind of microscopic vibration or movement beneath the threshold of perception.[47]

It cannot be denied that the plots of *Emma* and *Mansfield Park*, at least, seem concerned with ranking and placing, with putting everything in its place in an unchanging hierarchical order. Fanny Price's home in Portsmouth is "the abode of noise, disorder, and impropriety. Nobody was in their right place, nothing was done as it ought to be" (361). Or in *Emma*, the plot appears to establish a ranked order, with each person manifesting their own nature and finding their true level, with Emma and Knightley at the top, Harriet and Mr. Martin at the bottom. But in both cases, this unchanging order with its fixed positions is an object of merely personal desire, not shared by other characters, nor by the larger world of the novel. The ideal order of Mansfield Park exists only in Fanny's memory, or rather, rises up in her mind as the negation of her present surroundings: "She could think of nothing but Mansfield. . . . The elegance, propriety, regularity, harmony . . . were brought to her remembrance every hour of the day, by the prevalence of every thing opposite to them *here*" (363, original emphasis). It's a fantasy Mansfield, the product of Fanny's peculiar "frame and temper" (363). It is not the place or household that the novel actually represents. In *Emma* as well, we notice the remarkable freedom of movement and mixture that the plot enjoys and allows the characters to enjoy. Within the lava-like flow of Miss Bates's discourse, everything mingles; it is a language that Emma can also speak, "[flying] off, through half a sentence, to her mother's old petticoat" (177). Despite Emma's implicit belief in the rightness of a fixed hierarchy, the principles of the plot seem to be that anyone can go anywhere (thus the Perrys' need for a virtual carriage) and that nothing is too small to notice, or to become part of an event. The spatial and even social movement of characters within the given world is surprisingly unrestricted. Jane Fairfax, even though she is largely defined by her enigmatic immobility, is "seen wandering about the meadows, at some distance from Highbury" (308). Even Harriet's ascent to become the mistress of Donwell Abbey is "far, very far, from impossible" (325), Emma reflects with horror. It is not clear that everyone would share that horror.

The idea of a natural type becomes the object of satire. Emma invites Harriet to compare the manners of Mr. Martin with those of the "very real gentlemen" she has come into contact with: "At Hartfield you have had very good specimens of well educated, well bred men. . . . What say you to Mr. Weston and Mr. Elton? Compare Mr. Martin with either of them. Compare their manner of carrying themselves; of walking; of speaking; of being silent. You must see the difference" (26–27). Her presentation of the species, "marriageable male," echoes Buffon's logic: "There is in nature, a general prototype of each species, from which each individual is modelled, but which seems in procreation to be debased or improved, according to its circumstances; insomuch, that in relation to certain qualities, there is a strange variety in the appearance of individuals, and at the same time, a constant resemblance, which appears admirable, in the whole species."[48] But in the end plot leans toward "constant resemblance" and commensurability. There turns out to be between Emma and Mr. Knightley "equal worth. . . . No sacrifice on any side worth the name" (365, 368). The weighing is exact, and the time is noted scrupulously: "*Her* change was equal.—This one half hour had given to each the same precious certainty, had cleared from each the same degree of ignorance, jealousy, or distrust" (339–40, original emphasis). Concepts of inherent superiority live chiefly inside Emma's head.[49] The novel playfully equivocates with firstness as a position. It is not that Mr. Knightley is objectively "infinitely superior" but that he happened to come first in time, or first in the plot: "There had never been a time when she did not consider Mr. Knightly as infinitely the superior" (324), compared with Frank Churchill. Questions of type and rank are subsumed within a character's consciousness; the larger plots do not rely on them as an objective structure.

It is not exactly a chaotic world that Austen's plots invoke, but it is one in which motion is constant, affects every kind of body, and can be transmitted across substances and scales, even in novels apparently preoccupied with ordering. Portsmouth is a striking example of what the novel suggests might be a more real world: far from being a vacuum, the absence of "wealth and plenty" that Sir Thomas envisions, it is hypercharged with plot, with the inertial movement of bodies and the rearrangement of composite bodies, falling, tumbling, ascending, departing, swirling noisily: "The three boys all burst into the room together and sat down . . . they were still kicking each other's shins, and hallooing out at sudden starts" (355). Ships are always moving or about to move, the house itself moves ("The doors were in constant banging, the stairs were never at rest" [364]), particles of food adhere to plates and silverware. At the most dramatic points in the plots of both novels, there seems to occur a kind of leap or transduction across scales of being. Toward the end of *Mansfield Park*, Fanny sits in her family's house "in a blaze of oppressive heat, in a cloud of moving dust," and seems almost to dissolve into the small material dramas of inertial movement surrounding her: "The milk a mixture of motes floating in thin blue, and the bread and butter growing every minute more greasy. . . . Her father read his newspaper, and her mother lamented over the ragged carpet" (408). Like Brown's grains of pollen, to which they were compared, the particles in the dust move randomly; the "motes floating in thin blue" seem even closer to the "molecular reality" of ceaseless motion that Perrin described. When the news of Henry's and Maria's elopement erupts, it seems oddly continuous with the action of this scene, in which the smallest pieces are moving, mixing, adhering, and coming apart. As she herself passes "from hot fits of fever to cold," Fanny reflects with horror on how "intimate" this event is (410). Intimate though it is, the consequences of this event will continue to spread: "Whom would it not injure? Whose views might it not affect?" (410).

In *Emma*, a comparable moment of shock occurs at Box Hill, the event ruled by a "principle of separation." Rebuked by Mr. Knightley for a mean-spirited joke at Miss Bates's expense, Emma undergoes a dramatic change as she goes home: "The horses were in motion . . . with . . . unusual speed, they were half way down the hill, and every thing left far behind. . . . Never had she felt so agitated. . . . Emma felt the tears running down her cheeks . . . without being at any trouble to check them" (296). Here, a single agitation seems to run through the varied components of the scene. Moreover, the downward rush of horses and tears materially reenacts the social macrotrajectory of Miss Bates that's just been hammered into Emma's head by Mr. Knightley's reproach: "She has sunk from the comforts she was born to, and, if she live to old age, must probably sink more" (295).

If plot always has to have a clock in it, as E. M. Forster laments, Austen places that clock inside substances.[50] Characters, assemblages, and configurations are thought of as substances that possess a certain limited duration. The plot seems to pose this question at every turn: *how long will it last*? How long will the supply of apples, the piece of pork, the wedding cake, Harriet's love for Mr. Elton, Jane's stay in Highbury, Mr. Woodhouse's alarm about the turkey-burglar, Emma's delusions about her independence last? The answer depends on the material constitution of the stuff involved: its internal texture, arrangement, and quantity. Henry Tilney in *Northanger Abbey* says of Catherine's muslin gown, "It is very pretty . . . but I do not think it will wash well; I am afraid it will fray" (17). Emma's mocking imitation of Miss Bates is brilliantly on target here: "Not that it was such a very old petticoat either—*for still it would last a great while*—and indeed, she must thankfully say that their petticoats were all very strong" (177, emphasis added). Here is Miss Bates herself on how long the apples last: "There never was such a keeping apple anywhere as one of his trees . . . he asked whether we were not got to the end of our stock" (187).

It's not just a matter of Miss Bates's straitened circumstances. Emma notes with dismay how long Harriet's surprisingly retentive brain keeps the impression made by Mr. Elton: "The charm of an object to occupy the many vacancies of Harriet's mind was not to be talked away. [Mr. Elton] might be superseded by another . . . but nothing else, [Emma] feared, would cure her" (144). Harriet's mind, indeed, is a material arrangement that Emma observes with the detachment of a scientist. Emma notices the amount of contact necessary ("*just* to hear his voice, or see his shoulder") "to preserve [Mr. Elton] in her fancy," and the way that different objects fill those vacancies: "Sometimes Mr. Elton predominated, sometimes the Martins; and each was occasionally useful as a check to the other" (144–45). The model is mechanical; the problem is essentially one of time and duration: how long Emma will have to listen to Harriet's obsession with Elton? Volume 2 opens with this problem. Emma is "getting rid of the subject . . . but it burst out again when she thought she had succeeded . . . she found something else must be done. . . . She determined to call upon [Mrs. and Miss Bates] and seek safety in numbers" (123). But numbers and durations still pursue them: first "they must settle how long [Mr. Elton] had been gone," then how long Jane's letter is (two pages), how long until her arrival ("'Are you expecting Miss Fairfax here soon?' 'Oh yes; next week. . . . Yes, next week. . . . Yes, Friday or Saturday. . . . Oh, yes, Friday or Saturday next. That is what she writes about'" [124]). When Emma takes Harriet to visit the Martins as a desperate last resort, the question of duration becomes of paramount importance: how long since Harriet had been there, how long she stayed then, how long she stays now, and how this will set her old feelings into motion.[51]

Austen's plots entertain an idea of objectivity as the removal of the self, as the next chapter will explore. But her plots suggest another sense of it based on the texture of material bodies, considered as aggregations and intimate coalitions. Fanny's mind retains its impressions tenaciously: at Sotherton, "A quarter of an hour, twenty minutes, passed away, and Fanny was still thinking of Edmund, Miss Crawford, and herself" (91). Isabella Thorpe in *Northanger Abbey* is the opposite: "The mess-room will drink Isabella Thorpe for a fortnight, and she will laugh with your brother over poor Tilney's passion for a month," Henry tells Catherine, whereupon she "would contend no longer against comfort" (111). This measurement of time, the rate of change, seems particularly important to Austen's sense of reality, and it's where she most openly names the challenge of realist representation: "I purposely abstain from dates on this occasion. . . . I only intreat every body to believe that exactly at the time when it was quite natural that it should be so, and not a week earlier, Edmund did cease to care about Miss Crawford" (454). We do want to know: how long did it take, how many days, weeks, or months? The most moving discussion, however, comes deep within the fictional scene in *Persuasion* when Wentworth (not Austen) drops the pen. Fittingly, for a novel so concerned with the quality of "elasticity" as something more than metaphor, Anne and Captain Harville discuss the texture and toughness of feelings. Harville asserts, "As our bodies are the strongest, so are our feelings, capable of bearing most rough usage, and riding out the heaviest weather" to which Anne replies, "Your feelings may be the strongest . . . but . . . ours are the most tender." With a final contrast between "more robust" and "longer-lived" substances, she wins the debate (187–88). The final description of Anne's character seems carefully chosen: "Her character was now fixed on his mind as perfection itself, maintaining the loveliest medium of fortitude and gentleness" (194). The substance has been fixed at just the correct degree of ductility, equally supple and strong. It signals the artificial end of plot's real movements.

CHAPTER 7

Quixotism, Plot, and the Emergence of Mechanical Objectivity

Lady Delacour's disease is her secret, but not in the way that you might expect. In Maria Edgeworth's 1801 *Belinda*, mentioned admiringly in Austen's *Northanger Abbey*, this character believes she has breast cancer. Her enviable life is simply a performance that hides illness and domestic misery. But even when she confides in Belinda and reveals to her the "hideous spectacle" of her breast, Lady Delacour stresses "that which you cannot see—my mind is eaten away like my body, by uncurable disease—inveterate remorse," specifically, by shame about her shortcomings as a wife and mother.[1] Two things will help her: science and the plot itself. When she finally agrees to be operated on by a legitimate surgeon, we learn that it was nothing more than a painful bruise exaggerated by the unscrupulous quack who had tended her in secret. The disease consisted in her refusal to acknowledge to others her sense of guilt. She becomes a devoted mother and grateful wife. But the twist is that she could not have shown that inner truth deliberately or changed it by herself; only a dispassionate third party or instrument could reveal the objective truth of her mind's distress and her body's soundness—a surgeon's hand, the novel's plot.

The preceding chapters of this book have not considered in detail the role of human agents in plots as subjects who wish, hope, believe, know, or possess a unique kind of interiority.[2] This delay was deliberate, and reflects the historical dimension of this book's argument. In the current chapter, I will examine a change or rupture that happens toward the end of the eighteenth century in the way that knowledge is conceived. In simplified form, the argument is that subjectivity comes to be understood differently and to pose a problem for knowledge; earlier, it had been the unseen ground of representing and thus of knowing the world. This change does not happen all at once, however, or in a uniform way. But one way that we can observe it is by attending to the plots of novels written in this period. Do they show a shift in how humans regard themselves in relation to the knowledge they create? Do they demonstrate a wariness about subjective bias or self-absorption? Do they become aware of their own power as apparently objective instruments that can reveal hidden truths and create sustained interest in outcomes? In the course of taking up these questions, I will examine the novel's relation to science in a different manner. Instead of focusing on a branch of natural philosophy or a type of investigation, I'll look at one figure to see what it suggests about the dangers of subjectivity as well as the pursuit of science in this transitional period. That figure will be Sir Humphry Davy, a celebrated scientist with ties to Maria Edgeworth, her circle, and other well-known Romantic writers. Starting with the pedagogical narratives of Edgeworth and Anna Laetitia Barbauld, their science writing for children, we will see how the principle of plot acquires a greater independence and clearer function as something that can explore subjectivity, expose its hidden depths, and indicate its dangers and limitations.

In the novels examined here, plot comes to be shadowed by a certain figure whom Michel Foucault locates at the boundary between one regime of knowledge and another: Don Quixote.[3] For Foucault, the hapless Don Quixote marks the end of the Renaissance episteme, in which resemblances between things were accepted as a legitimate form of knowledge. Cervantes's figure, he argues, inaugurates the classical regime of knowledge in which "what has become important is no longer resemblances but identities and differences," or, in other words, the kind of knowledge that structures many of the plots we have examined so far.[4] It is apt, then, that quixotism should return as a kind of transitional object for the novel as that episteme is superseded by a modern one based on assumptions about depth and interiority.[5] I will suggest that quixotism, with its long latency in the English novel, bears a particular relevance for the genre at this point because of several related features that now emerge more clearly. Quixotism ties the shape of the plot to the distorted beliefs and perceptions of its protagonist. Although Cervantes's novel features many characters and interpolated stories, it is the unstoppable, unchangeable nature of its protagonist's subjectivity that accounts for the overall shape of its plot, as far as it has one, and this was the case with any type of quixote narrative, including the female one. Plots can now emulate this drive while angling it to reveal the inherent limitations of subjective knowledge rather than the delusions of one character. Plot flaunts its own power as an instrument of discovery, bringing its energies to uncover supposedly objective truths about things, events, and human subjects that cannot otherwise be properly seen. I will suggest that we can even see some resemblances between the principle of plot and what Lorraine Daston and Peter Galison have called mechanical objectivity, a kind of objectivity produced by carefully eliminating subjective bias and interference.[6] Plots claim to reveal the contents of strange, inaccessible, interior spaces that cannot otherwise be known.[7]

A special kind of impersonal or objective suspense also comes to be attached to plots when they incorporate quixotism. In its earlier iterations, quixotism did not emphasize suspense as a global feature of plot. This is not to say that there isn't surprise in Cervantes's novel, or in the many imitations that followed, but that quixotism in England in the eighteenth century was a known quantity, a topos or commonplace theme.[8] As a result, surprise lay more in the surface variations that the plot could orchestrate, than in the question of what will happen or what will come to light. Later in the eighteenth century, though, quixotism becomes an unexpected source of suspense in its own right.[9] And suspense starts to be understood as something that can counteract a tendency to remain trapped within the self. Classic quixotes are those who cannot feel suspense because they are confident that they already know how everything will turn out. Plots can now draw quixotes out of themselves, and draw us out of ourselves by hinting at secrets but not fully unveiling them until the end, and maybe not even then. If plots had aimed to show things, to trace intermediate causes, to manipulate and compare bodies or substances, they seem equally happy now to hide things, to hoard their own knowledge. This is, of course, a venerable feature of plot; Aristotle had praised the plot that turns on sudden, surprising recognitions.[10] Such recognitions can be found in Fielding's novels, and in countless others.[11] But now the ability of plot to produce and manage knowledge takes on a new efficacy. We are invited to think of plot as directly embodying a sustained attention to continuous processes of disclosure that unfold independently of anyone's wishes. The suspense that plot thus generates will be viewed as having its own therapeutic benefits, when properly handled and applied. It can encompass long sequences of events or attach itself to small scenes, little things and their microplots. But, importantly, suspense now appears to belong to plot itself as an objective process of development or disclosure. Instead of being a matter of subjective curiosity or surprise, suspense tracks or registers the gradual unfolding of knowledge outside you.

In the plots of novels considered in previous chapters, the varied subjectivity of characters did not appear to pose a special challenge to representation: ultimately, they could be shown and compared to each other, or rather, they could not *not* be shown through what they did, said, wrote, how they moved, reacted. Plot traced intermediate causes, ordered natural types, observed and investigated changes on an intimate level, experimented with aggregation and disaggregation, all in ways that involved humans but did not see subjectivity as a threat to a larger project of showing and thereby knowing. Even when characters were driven by will, desire, or imagination, even when they were as eccentric as those in *Tristram Shandy*, these phenomena were part of their representable and therefore knowable being. But there is broad consensus from thinkers as diverse as Georg Lukács, J. G. A. Pocock, Louis Althusser, and Foucault that an important shift occurs toward the end of the eighteenth century in how knowledge was conceived in relation to subjectivity.[12] As we saw in the fourth chapter, the eighteenth century, according to Foucault, identified knowledge with that which was made visible to the eye and ordered in terms of identities and differences, and it did so without fundamentally questioning how such ordering or visibility depended on the human viewer or whether there were essences or relations that eluded such representation. The goal of natural knowledge was to arrange empirical entities (plants, minerals, animals, wealth, words) into an "articulated system . . . a table on which knowledge is displayed."[13] Language smoothly articulated itself with other systems and could thus represent everything: "The fundamental task of Classical 'discourse' is to ascribe a name to things, and in that name to name their being."[14] Taxonomies instantiated this way of seeing. In the world, taxonomists saw "lines, surfaces, forms, reliefs . . . the extension of which all natural beings are constituted . . . the form of the elements, the quantity of those elements, the manner in which they are distributed in space in relation to each other, and the relative magnitude of each element."[15] In natural history, Linnaeus asserted that we only needed to notice the number, form, and disposition of certain clearly visible parts in plants and animals, "variables of description . . . which could be scanned . . . by language and by the eye."[16] While other natural historians such as Buffon deplored the narrowness of the criteria used by Linnaeus, they also accepted the eye's ability to see and language's capacity to name things and articulate their relations. Classical nature was a "homogeneous space of orderable identities and differences . . . a unitary field of visibility and order."[17]

In Fielding, Richardson, Smollett, and Sterne, the representability of character was axiomatic, even if, as in Sterne, there was a struggle to confine it within a linear narrative order. It continued to remain an aspiration in Burney, though an increasingly challenging one, to make the substance of character fully apparent to the observing eye. Even with all their biases, self-love, and self-deception, characters manifested themselves in the representational field or plane of the plot through their own words and actions—even (or especially) when they sought to disguise themselves. Even when, as in *Clarissa*, characters depend on their dialectical opponent to learn who or what they are, it is not taken for granted that one point of view possesses greater objectivity than another. The sequence of Clarissa's letters could not fail to represent her being to the reader, even if she could not always read her own letters correctly. Richardson's editor and Fielding's narrator are in a sense related to their texts somewhat like the royal couple in Velázquez's painting *Las Meninas*, who anchor the entire representation almost invisibly (dimly reflected only in the mirror) and stand in the place of the viewer, as Foucault argues.[18] Like a canvas, plot furnished a plane where a character's nature became visible spontaneously. Plot thus blended into discourse and description as an aspect of representation, and its task was to trace movements, to ascertain, order, and arrange things. If a character's substance was not clear, it was because not enough of it had been brought into view. More letters recounting what they said, thought, and did were required, or more dynamic encounters that would reveal what stuff they were made of, who was stronger, braver, more cunning, more innocent than another. The visible counted as knowledge; its objectivity had not been made a problem, nor had its reach or capacity.

Toward the end of the eighteenth century, this regime began to shift, according to Foucault. Instead of the surface differences important to classification, the new discipline of biology sought to understand the "hidden foundation" of the living organism, its "invisible focal unity."[19] Rather than tabulating "the possibilities of being" in all of the latter's empirical, fully accomplished diversity, natural philosophers sought to understand the connections between the organism's inner hierarchy of functions and "the conditions of life."[20] This shift opened up a realm of interiority that could not be directly viewed.[21] It could only be inferred from the anatomy and functioning of organisms. As Foucault puts it, "A profound historicity penetrates into the heart of things, isolates and defines them in their own coherence . . . seeking the principle of their intelligibility only in their own development, and abandoning the space of representation."[22] To know something, in other words, it was no longer enough to see its forms or the arrangement of its parts and to compare those to other beings. There was something deeper inside, something that shaped its form and drove its behavior.

In Georges Cuvier's "Preliminary Discourse" (1812), for example, the living creature's need to sustain its existence in its own unique way ties together even distal features.

> Every organized being forms a whole, a unique and closed system, in which all the parts correspond mutually, and contribute to the same definitive action . . . if the intestines of an animal are organized in such a way as to digest only flesh . . . it is also necessary that the jaws be constructed for devouring prey; the claws, for seizing and tearing it; the teeth, for cutting and dividing its flesh; the entire system of its locomotive organs, for pursuing and catching it; its sense organs, for detecting it from afar; and it is even necessary that nature should have placed in its brain the instinct necessary for knowing how to hide itself and set traps for its victims . . . every animal adapted . . . for this regime unfailingly combines them, for its species could not have subsisted without them.[23]

If it had been enough to count teeth and hooves and note their arrangements in order to relate animals to each other, the means and goal of knowledge look different now. It looks inward, into an invisible space, an "inaccessible point, which drives down, beyond our gaze, towards the very heart of things."[24] Cuvier posits a hidden source of unity that makes the animal or species what it is. There seems to be a kind of blind cooperation within the being that makes a "definitive action" possible. We can note that this description, "all the parts . . . contribute to the same definitive action," echoes a familiar Aristotelian account of plot as well as a familiar model of the literary text as a unified whole. But the unity is now located deep within the being rather than on its surface; it has to be inferred rather than seen, and it is explicitly connected to the struggle for being, the conditions of life as an abstract or universal concept. Cuvier's description of this animal vividly evokes a plot, a sequence of actions (hiding, setting traps, springing out, pursuing, catching) that shows what it is more truly than any static morphology can do. And those actions, in turn, are made necessary and possible by the conditions in which the creature has developed.

How, then, does plot respond to this challenge? Here we can turn to Daston and Galison's complementary account of the history of objectivity. They too mark the end of the eighteenth century as the moment of a shift in science's attitude toward the visible. They cite Linnaean classification as an example of "science before objectivity," or the earlier scientific ideal that they call "truth to nature."[25] In the eighteenth century, the makers of scientific images did much more than observe closely. They intervened, made judgments, synthesized, and made alterations in order to represent the ideal type of the object rather than singular haphazard specimens. "This knowing self is a precondition for knowledge, not an obstacle to it," Daston and Galison argue.[26] In this period, "idiosyncratic objects," more than idiosyncratic subjects, elicited the greatest concern.[27] This faith in the truth or the space of representation itself, as Foucault calls it, does not entirely disappear, but comes to be modified in the nineteenth century.[28] Now we start to see the familiar pairing of objectivity and subjectivity as mutually exclusive. In this newer conception of knowledge described by Daston and Galison, "Objectivity is the suppression of some aspect of the self, the countering of subjectivity. Objectivity and subjectivity define each other."[29] Thus, in the context of the scientific atlases that they examine, objectivity came to be defined in terms of the image that was produced with the least possible intervention of the human observer: "By mechanical objectivity we mean the insistent drive to repress the willful intervention of the artist-author, and to put in its stead a set of procedures that would . . . move nature to the page through a strict protocol, if not automatically. . . . These protocols aimed to let the specimen appear without that distortion characteristic of the observer's personal tastes, commitments, or ambitions."[30] I want to suggest that plot, in the realist novel, begins to assume something of the quality of mechanical objectivity. Through its protocols, plot itself could, with apparent objectivity, elicit causes that are assumed to be otherwise hidden, inaccessible. If plot had been used to emphasize intermediate causes, the nexus of things that allowed events to happen, it does so now with a different appreciation of where the boundaries of knowledge may lie, of what is hidden and to whom. As opposed to what characters themselves might hope, fear, or wish, plot seems to hold out the promise of objective truth—but as a goal, rather than as something given by the act of representation itself. Plot can show things without showing them fully. Sometimes, as in Edgeworth, elaborate procedures are invented by the plot in order to disclose someone's secret truth, a truth that we know is there but have not yet seen. Experiments, accidents, and instruments can all play a role. Daston and Galison define mechanical objectivity as "a determined effort to see without the distortions induced by authority, aesthetic pleasure, or self-love"; it relies on "external procedures" as well as active self-suppression.[31] But plot has no self to suppress. Thus, it can pursue and unveil. It also becomes axiomatic in the novels considered here that characters cannot intentionally display their innermost truths without succumbing to the lie of theatricality.[32] Such disclosures are still too attached to their own wills; they are too subjective. Lady Delacour, for example, cannot see or disclose herself truly, but another character sees the shadow of a collar around her neck vibrating with a feverish pulse. Ultimately, the procedures of plot will perform multiple disclosures, and provide a cure for what appears to be an endemic (or pandemic) quixotism.

Daston and Galison attribute this shift at the end of the eighteenth century to the way that Enlightenment psychology's model of a passive subject was replaced by an active Kantian one, "fusing raw sensations into coherent experience. Organized around the dynamic and autonomous will, the self acted on the world, projecting itself outward."[33] But there also existed a more familiar, pervasive model of a subjectivity that distorts the external world and projects itself onto it in disastrous ways. Moreover, it is found in a work that came to be seen, from around this time, as the quintessential or paradigmatic novel.[34] The topos of quixotism now becomes incorporated into a general idea of subjectivity.[35] Cervantes's novel, *Don Quixote* (1605, 1615), was translated into English in 1612, and its second part in 1620 (both by Thomas Shelton). It was repeatedly retranslated over the course of the eighteenth century. The abstraction, "quixotism," was already used by 1664, according to the OED. Henry Fielding's play *Don Quixote in England* (1734) imports the characters into England and places them in the familiar territory of Restoration comedy, where they seem perfectly comfortable.[36] While the figure of the quixote occurs in English novels throughout the eighteenth century, I want to suggest that the problem of quixotism takes on a renewed relevance in this period of transition. Quixotism offers an apt model for what became the scientist's enemy by the end of this period: "The subjective self . . . was viewed *as overactive and prone to impose its preconceptions*."[37] It blends into the general idea of subjectivity, and becomes part of the texture of the novel, the warp of its plot.

This required redefining as well as normalizing the existing idea of quixotism. Both natural philosophy and the novel appear newly intrigued by the problem and the implicit solutions that they could offer. The problem is twofold. Subjectivity has a tendency not only to distort the perception of the external world but also to implode, to fall in on itself; quixotic plots thus have the tendency simply to repeat versions of the same encounter again and again. The solution, for novels, lies in the way that plot can elicit truths objectively, without anyone's active participation. But both natural philosophy and the novel also seem aware of the question of what we might call plot continuity. Instead of the serial repetition of "adventures" or accumulation of episodes, novels seem to recognize afresh how suspense can be created through delayed, overlapping rhythms of causality situated in the external world. Edgeworth and other novelists start to conceive of plot as the continuous, objective, and suspenseful disclosure of interior causes that are hidden within nature or reality.[38]

While a full-scale consideration of quixotism is beyond the scope of this chapter, two examples from earlier in the eighteenth century can provide touchstones for comparison.[39] In Henry Fielding's *Joseph Andrews*, "Written in Imitation of The Manner of Cervantes, Author of Don Quixote," Abraham Adams is one type offered up among an array of others. His quixotism consists in a kind of book-fetishism: both an attachment to a particular physical book (referred to simply as "Aeschylus") and a more general belief that books have greater truth-value than experience. Indeed, his attachment to books seems to inoculate him from experience. He cannot change, and he cannot learn, despite numerous situations that reveal unmistakably to everyone else (if they didn't know already) that "passions such as malice and envy exist in mankind."[40] He is "ignorant of the ways of this world" (18), and it is not questioned that this ignorance, instead of being overcome, will be displayed again and again for our amusement in the form of serially repeated adventures.[41] As Sarah Fielding and Jane Collier put it in *The Cry*, "The noble simplicity of his mind with the innumerable beauties in his character . . . are placed in the most conspicuous view."[42] Conspicuous, indeed: a typical chapter (book 2, chapter 16) is titled, "A very curious Adventure, in which Mr. Adams gave a much greater Instance of the honest Simplicity of his Heart than of his Experience in the Ways of this World" (134). Fielding's novel relies on "variables of description . . . which could be scanned . . . by language and by the eye," and his plot seems to rely on repetition to further stabilize the representation of nature.[43] As an overarching trajectory, plot serves to confirm what we already know and what we've already seen. Regardless of the diversity of mistakes, discoveries, and accidents, each event reaffirms that Adams is and could be nothing other than what he is—a combination of Don Quixote's high-minded idealism and Sancho Panza's naïveté. Objectivity, as I argued in the fourth chapter, exists on the representational plane of the plot. Plot is that canvas. It is not possible that any turn of events could bring Adams to understand himself better, for there is nothing deeper to understand.

A more intricate relation of quixotism to plot emerges in Charlotte Lennox's *The Female Quixote* (1752), a work that was included in Barbauld's 1810 *British Novelists* but excluded from Scott's *Ballantyne's Novelist's Library* in the early 1820s.[44] In the case of Arabella, its protagonist, quixotism is represented not only as an immunity to experience, but as resistance on her part to a continuous developmental plot. Having internalized seventeenth-century French romances, but also in some ways showing a very Newtonian confidence, Arabella sees the external world as regular, consistent, and ruled by causal laws—only these laws are drawn entirely from the fiction she has read. Not only are there "Rules of Heroick Virtue" to follow, there are also laws about how things happen, and thus the same sequences are bound to recur, in Arabella's view. For example, beauty is fatal to those who see it and therefore fall in love with it: "'Tis very certain, my Beauty has produced very deplorable Effects . . . by a fatal Necessity, all these Things will happen, whether I would or no. . . . Questionless [Sir George] must die . . . if he persists in his Design of loving me" (201). And this is bound to happen any number of times. Other characters are amazed at her "facility in accommodating every incident to her own wishes and conceptions" (39). Glanville, who wishes to marry her, cannot do so "till the Whims her Romances had put into her Head were eraz'd by a better Knowledge of Life and Manners. But . . . he knew not how this Reformation would be effected; for she had such a strange Facility in reconciling every Incident to her own fantastick Ideas, that *every new Object added Strength to the fatal Deception she labored under*" (380, emphasis added). Objects or even objectivity cannot cure her, for the more she attends to the external world, the more she sees what she always saw. Her boundless faith in the objective operation of universal laws provides the exact image of her problematic subjectivity, or, in the novel's term, "singularity."

The formal problem and the epistemological one are two aspects of the same thing. The plot can only repeat versions of the same episode because Arabella's quixotism won't allow her to see anything other than the same universal laws realizing themselves. But Lennox's novel seems to balance two paradigms of plot against each other: the iterative quixotic plot, ruled by the repetition of discrete, clearly bounded and highly similar adventures, and the novelistic plot, which traces change and extends over the whole narrative, though in this case it does take an abrupt and somewhat awkward turn at the end. In the quixote's view, plot is the always-enthralling repetition of a "great number of cares, disappointments, and distresses of various kinds" (41). Arabella states, "I believe there are few young Ladies in the World, who have any Pretensions to Beauty, that have not given Rise to a great many Adventures; and some of them haply very fatal" (108). This is her understanding of experience, and to a certain degree the novel also presents itself in this fashion. But against this, the novel sets the more continuously graded and apparently open-ended plot of Glanville's embarrassment at her eccentricity—an embarrassment that repeats itself but also rises to desperation as Arabella becomes exposed to larger social circles. He says to another character, "Lady Bella's Peculiarity . . . gives me a real Pain . . . I must cure her of that Singularity; and therefore I beg you . . . help me to banish them from her Imagination" (225).

Arabella's quixotism, like Adams's, cannot be cured by experience and cannot really be cured by the events orchestrated by the plot. It falls to a long dialogue at the end of the novel, in which an unnamed "pious and learned doctor" tries to change her mind through a prolonged debate. In it, Arabella shrewdly identifies all the epistemological weaknesses of empiricism. Although the doctor tries to dismiss Arabella's cogent objections, she seems to have the upper hand in the argument until he bluntly asserts his superior epistemic authority on the basis of his age, gender, and experience (ignoring Arabella's critique of the latter):

> Your Ladyship must suffer me to decide, in some Measure authoritatively, whether Life is truly described in those Books; the Likeness of a Picture can only be determined by a Knowledge of the Original. You have yet had little Opportunity of knowing the Ways of Mankind, which cannot be learned but from Experience, and of which the highest Understanding, and the lowest, must enter the World in equal Ignorance. I have lived long in a public Character . . . I therefore presume to tell our Ladyship . . . that your Writers have instituted a World of their own. (423)

Those writers' accounts of the world and its laws are merely subjective fantasies rather than true representations, he insists. She has no choice but to take his word for it, and so she does. This conclusion is less an experience she might undergo than a wholesale hijacking of experience. Lennox's novel, like Fielding's, does not allow for the possibility that other kinds of objective events in the plot could give rise to a different understanding of the world or of oneself.[45] The more intensely you scrutinize the world, the more you will see in it only events that confirm the rules of your quixotic universe. In both of these novels, the quixote is anomalous rather than typical of its represented universe. The plot can only accommodate one, and other characters are sorted by their respective relations to her or him (friend, accomplice, abuser, disabuser, etc.).

But by around 1800 quixotism seems to reflect a more common condition. Edgeworth's *Belinda* displays a remarkable range of quixotic pathologies and their cures. Quixotism becomes internalized and endemic. But instead of leading Lady Delacour and other characters to mistake the identity of things in the external world, taking windmills for giants, for example, quixotism is likeliest to occur when it comes to perceiving, or rather, failing to perceive oneself. It does not always issue from reading romances—although for one character in Belinda it does come about through reading Rousseau's *Emile*—and it does not always ensue in large, dramatic misprisions. More often it takes the form of a persistent, willful, dangerous bias or partial blindness. In other words, it merges with the concept of subjectivity, which takes on the coloring of quixotism. Interestingly, such quixotic self-absorption could be cured by a sustained focus on external phenomena as the latter play out over time. Thus, plot can benefit from quixotism as a problem to solve (or alleviate) at several levels.[46] To the extent that its procedures rely on external objects to elicit the subject's truth, we can think of such plots as a kind of successor to the original quixotism of Cervantes, whose novel also exhibits the reasoning and behavior of its anomalous protagonist. But instead of satire, these later plots invite suspenseful observation, a type of experience that is meant both to disclose and to counteract quixotism.[47]

I want to turn now to the figure of Humphry Davy, who can be located at the intersection of multiple historical transitions. His career links older institutions such as the Royal Society with new, more public-oriented ones like the Royal Institution, and also stands between an older model of natural philosophy and the emergence of more specialized scientific disciplines.[48] Davy came to be seen as one of the leading scientific discoverers of his age, and was also perhaps the first celebrity scientist, as Jan Golinski has shown. Building on a growing public interest in science, Davy drew on his abilities as a charismatic lecturer to attain an unprecedented popularity. Golinski argues that it was not only the subject matter or scope of Davy's scientific discoveries but his self-presentation as a genius that attracted such public enthusiasm. It reflected the new attention that came to be given to the natural philosopher as an individual subject. While Newton's laws were universally known, Davy's fame rested on the centrality and visibility of his own body in a very different way. Golinski argues that, despite his numerous discoveries, including those of many chemical elements, Davy did not exert lasting influence on institutional forms of science. His influence "remained tied to Davy's embodied persona," and his attitude toward natural philosophy would soon become outdated.[49] While Davy argued that "the man of true genius who studies science" will "combine together mechanical, chemical, and physiological knowledge," that view was not widely shared; specialized scientific societies were established during the course of his career, as well as the British Association for the Advancement of Science.[50] We can also see that Davy's practices and assumptions did not fit very well an emerging model of mechanical objectivity. He seems aware of this; a certain acknowledgment of quixotism, of the self as a potential problem, can be seen in his published researches.

From the outset, Davy's reputation as an "experimental chemist" was built on his willingness to use himself as an experimental subject. His *Researches, Chemical and Philosophical; Chiefly Concerning Nitrous Oxide, or Dephlogisticated Nitrous Air, and Its Respiration* (1800) begin with systematic experiments on various animals. He comes to the conclusion, though, that he has to breathe it himself, "to inspire it in its pure form, for I saw no other way in which its respirability, or powers could be determined."[51] He measures the quantity of air he can inhale, chemically analyzes his exhalations, and documents the effects on himself of breathing various gases. The fourth section includes a journal in which he notes his sensations as he continues to experiment with breathing nitrous oxide over the course of around fourteen months. He seems to recognize the problem of distinguishing his own subjectivity from the effects of the gas itself. His second trial in April 1799 fills him with a "thrilling" sensation: "The objects around me became dazzling," and he is filled with what he calls "an irresistible propensity to action." But the following morning, he is filled with doubt: if he had not written it down, "I should even have doubted of their reality. I was willing indeed to attribute some of the strong emotion to the enthusiasm, which I supposed must have been necessarily connected with the perception of agreeable feelings, when I was prepared to experience painful sensations. Two experiments however, made in the course of this day . . . convinced me that the effects were solely owing to the specific operation of the gas."[52] Davy tries his best to separate the possible psychological causes of physiological sensations from those owing to the gas: "The general effects of its operation upon my health and state of mind, are extremely difficult of description; nor can I well discriminate between its agency and that of other physical and moral causes. . . . I had a constant desire of action, a restlessness, and an uneasy feeling about the praecordia analogous to the sickness of hope. But perhaps these phaenomena in some measure depended on the interest and labour connected with the experimental investigation."[53]

Davy's experiments with nitrous oxide read almost like journeys into a pure, enclosed interior space of subjectivity. Each time he finds that after breathing it, "impressions ceased to be perceived. . . . I gradually began to lose the perception of external things."[54] At one point he enters into an air-tight "breathing box," into which nitrous oxide is pumped, and remains there for more than an hour. After emerging, he notes, "I lost all connection with external things. . . . I existed in a world of newly connected and newly modified ideas." Famously, he exclaims to his gathered witnesses, "Nothing exists but thoughts!—the universe is composed of impressions, ideas, pleasures and pains!"[55] It is not too difficult to see the scientist as quixote here. When he comes out of the box, his first emotions are the characteristic quixotic emotions of "indignation and pride . . . produced by the sight of the persons about me. My emotions were enthusiastic and sublime." And one of the most consistent effects produced on or in him by the gas is "that feeling of restless energy, or that desire of action connected with no definite object, which I had often experienced in the course of experiments," a feeling quintessentially novelistic, according to Lukács's later theory of the genre.[56] It is not surprising to find Davy writing in 1800, "I have been puzzling myself to find out what people mean by external things."[57] The issue is not so much his singularity, what makes him similar to or different from the many other people who visited the Medical Pneumatic Institution, but rather the connection between his inner experiences and the external world.[58]

In 1801, Davy moved to the Royal Institution in London, where he performed dramatic experiments, often involving the use of his own body, to a lecture hall with hundreds of spectators.[59] In his "Discourse Introductory to a Course of Lectures on Chemistry, Delivered in the Theatre of the Royal Institution on the 21st of Jan, 1802," he places chemistry at the center of a map of natural philosophy:

> Mechanical philosophy, regarded as the science of the motions of the masses of matter . . . is . . . dependent upon chemical laws. . . . It is to

> chemistry that he is indebted for the knowledge of the nature and properties of the substances he employs. . . . Natural history and chemistry are attached to each other by very intimate ties . . . the first of these sciences treats of the general external properties of bodies, the last unfolds their internal constitution and ascertains their intimate nature. Natural history examines the beings and substances of the external world, chiefly in their permanent and unchanging forms; whereas chemistry, by studying them in the laws of their alterations, develops and explains their active powers and the exertions of those powers.[60]

In the division of labor between natural history and chemistry that he describes, the former is occupied with external and unchanging forms; the implication is that natural history ensues in collections. Chemistry, in contrast, studies how things change and how they interact at an "intimate" level; the implication is that chemistry deals with plots, that it handles the same material as does the novel. His language even approaches that in which the genre of the novel was beginning to be described: "Chemistry is that part of natural philosophy which relates to *those intimate actions of bodies upon each other, by which their appearances are altered, and their individuality destroyed*. . . . This science . . . relates not only to *the minute alterations in the external world, which are daily coming under the cognizance of our senses* . . . but likewise to the great changes, and convulsions in nature, which, occurring but seldom, excite our curiosity, or awaken our astonishment."[61] A reader familiar with Austen's *Persuasion* might think of the phrase, "So altered that he should not have known you again," as it is reported by Mary, echoed silently by Anne several times, and then played again in Wentworth's head.[62] It also echoes the account Clara Reeve had given of the genre in her work, *The Progress of Romance* (1785): the novel concerns itself with "such things as pass every day before our eyes."[63] Reviewing Austen's *Emma*, Walter Scott would reiterate that the novel concerns itself with "that which is daily taking place around [the reader]."[64] Both the novel and the science of chemistry as Davy describes it concern themselves with quotidian, intimate alterations that bodies bring about in each other.

Given his tendency in his early research to lose himself in his own interiority, we can find it ironic that Davy praises chemistry not only for its "great agency upon the improvement of society" but also for how it provides "attachment" to the external world. For people "in need of sources of permanent attachment, the cultivation of chemistry and the physical sciences may be eminently beneficial," he argues: chemistry "may attach the affections to objects, permanent, important, and intimately related to the interests of the human species."[65] This remark seems to be aimed at upper-class people suffering from "ennui" of the kind that Edgeworth describes in her novel by that name: the affluent, chronically bored, and listless. But Davy seems to target a condition more specific than boredom, that is, a tendency to become trapped within one's own fixed ideas: "The contemplation of the various phaenomena in the external world is eminently fitted for giving a permanent and placid enjoyment to the mind. For the relations of these phaenomena are perpetually changing; and consequently they are uniformly obliging us *to alter our modes of thinking*."[66] These external phenomena are therapeutic because they are outside you, and also because they are "perpetually changing," involved in an ongoing plot. Davy stresses that scientific research generates a kind of suspense, a sense of temporal unfolding that produces in his view a beneficial cognitive and affective engagement. The uncultivated person "[lives] only in moments, calculates but little on futurity. He has no vivid feelings of hope, or thoughts of permanent and powerful action. And unable to discover causes, he is . . . passively submissive." But by engaging in science, examining external things, and thinking about causes, a person becomes "capable of anticipating." Chemistry can "demonstrate that every being is intended for some definite end or purpose"; every being, in other words, has its place in a plot.[67] In his account of chemistry as a pursuit, Davy seems to anticipate the physiological study of novel-reading that Nicholas Dames has examined. That interdisciplinary investigation, carried out largely in the second half of the nineteenth century, focused on the reader's nervous system, emphasized the continuous experience of reading, and drew on "a sense of plot as . . . 'pure contiguity': a one-thing-then-another sense of plot . . . a series of affective 'moments.'"[68]

That sense of plot as an urgent experience of contiguity is something we can find even before the elaborate multiplot works of Dickens, or sensation novels. The overwhelming need to find out what happens next was prominently associated with the genre of the Gothic novel. In Austen's *Northanger Abbey* (itself a notable example of quixotism), Henry Tilney describes the effect of Radcliffe's *Mysteries of Udolpho*: "When I had once begun it, I could not lay down again—I remember finishing it in two days—my hair standing on end the whole time. . . . Here was I, in my eagerness to get on, refusing to wait only five minutes for my sister . . . keeping her in suspense at a most interesting part, by running away with the volume."[69] Rather than the Gothic, however, I want to look at a lesser-known genre of narrative that bridges natural philosophy and the novel. While Davy's connection with Coleridge and the Romantic poets has been studied, his connections with the Edgeworth circle have not been as thoroughly explored, even though his original collaborator at the Medical Pneumatic Institution, Thomas Beddoes, was Maria Edgeworth's brother-in-law.[70] Edgeworth and Anna Laetitia Barbauld both wrote narratives for children that not only disseminated scientific ideas but also recommended exactly the kind of cognitive and affective engagement with the external world that could result from this intersection of science and narrative. Their tales and scenes rely on plot and everyday suspense in ways that make an explicit connection to the problem of quixotism. Though Barbauld's and Edgeworth's writings in this genre were ridiculed by men like Coleridge and Lamb, their books were widely influential and remain worth reading for many reasons. They also resemble the novel in striking ways.[71] Frances Ferguson has recently pointed out that Barbauld, in the introduction to her *British Novelists*, "highlighted features of the genre that showed it to be a near relation to her writings for children."[72] Barbauld's *Evenings at Home*, for example, are often written in the form of dialogue: "What is significant about the scientific conversations is not so much that they yield up invariable factual statements. . . . Rather, those discussions point to the ways immediate experience and close observation of the natural world enable someone to take up a line of thought and continue it."[73] I would go further and suggest that these works, Edgeworth's *Early Lessons* in particular, form a bridge between the novel and natural philosophy at a moment when both were involved in this transition in the model of knowledge. Their simple narratives and dialogues offer lessons in natural history, Newtonian mechanics, even chemistry; they also try to teach self-discipline, a version of the kind of self-abnegation that was becoming central to the scientific ideal of objectivity. They recommend an attitude of attentiveness to the external world, and they use the unfolding of plot to stimulate it. A tendency to become trapped in one's own subjective feelings or thoughts is shown to be a natural condition, which the attractions of plot and science can counteract.

Edgeworth's *Harry and Lucy* began as a short two-part narrative written with her father and published in 1801 as part of *Early Lessons*. In 1825, she added a long sequel in multiple volumes. Thus the writing of this work spans a period in the history of the novel that includes the publication of all of her own novels, all of Austen's, and most of Sir Walter Scott's.[74] Her 1825 preface notes that the original *Harry and Lucy* "was, I believe, the very first attempt to give any correct elementary knowledge or taste for science in a narrative suited to the comprehension of children."[75] It arose from "what everybody now feels, that the taste for science, as well as literary knowledge, which has risen rapidly, and has spread widely, would render it necessary to make some provision for the early instruction of youth in science" (xii–xiii). But this work does more than provide a scientific education. It identifies quixotism as a constant threat that can best be overcome by cultivating an interest in how things unfold of their own accord, objectively, in the external world.

Edgeworth assumes that for children in particular, both attentive reading and close observation of the external world are motivated by suspense, by wanting to know what will happen next or as a result of something. In the 1825 preface, Edgeworth specifically addresses the question of plot or, as she calls it, "incident and story." Apologizing for its relative absence, she hopes that "the varying occurrences of domestic life, the frequent changes of scene, and the different characters of the children, with all their hopes and fears in the pursuit of their own little schemes and experiments, will . . . produce sufficient action to create interest" (ix–x). In some ways, the 1801 version, with its simpler characters, more successfully promotes a sense of plot, but the later sequel stresses more explicitly the joint emergence of quixotism as a problem and quasi-mechanical objectivity as an ideal.

The 1801 version gives "an account of three days passed by Harry and Lucy. One day when Harry was about five, and Lucy six years old; and two days a year afterward, when Lucy was seven, and Harry six years of age."[76] This lapse of time allows plot to emerge.[77] Harry, for example, makes a promise in the first part and fulfills it in the second part. In the course of daily experiments with their father, the children themselves stress the same suspenseful interest that Davy had recommended: "'I love to have something to do, and something to think of,' said Harry. 'And something to feel eager to go to again the next day,' said Lucy. 'I like to feel curious to know how the thing will turn out.'" (149–50). How the thing turns out (as well as how the children turn out) will depend not only on subjective choices and decisions, but on objective factors and the pursuit of objectivity. Anticipation and fulfillment rest on the behavior of natural phenomena in the external world. Plot is transferred to the objects the children examine: seeds in a plate, glass bottles filled with water, and so on. How will they behave? Will the frozen bottles break, will the seeds grow?

The sequel follows the children five or six years later; Harry is now eleven. The narrative, which is certainly long enough to be a novel (more than a thousand pages!), takes the children on a journey to the seaside with their parents, with stops along the way at cotton mills, iron foundries, sugar- and glass-works, and other places of scientific and industrial interest. Here, Edgeworth emphasizes the contrasting characters of the two children, or, as their father puts it, their "opposite faults."[78] Lucy's attention "passes too easily from one thing to another." She is constantly distracted by her own associations, by the fluttering and restless movements of her own mind. Harry, though, tends to become obsessed with his own ideas—steam engines, the sea, or electricity. His father notes, "You lost a whole month of your life, trying to make an electrical machine. I found your head so full of that book that I could get nothing else into it."[79] Harry himself admits that often "some fixed notion has taken possession of my wise head beforehand, and of which I cannot get rid . . . this prejudice prevented me from thinking of the right way."[80] Both the children, Harry in particular, become increasingly interested in trying to see their own minds and to view their own mental idiosyncrasies objectively. They refer to "the wonderful inside of our own minds" (312), and at one point they even read aloud from Locke the following passage: "'Men thus possessed in company'—'that is, with their own favourite thoughts,' said Lucy—'Are as if they are under the power of enchantment. They see not what passes before their eyes'"; they are trapped in "'their secret cabinet within, where they have been wholly taken up with the puppet which was for that time appointed for their entertainment'" (260–61). Allusions to Don Quixote and to "a romance called Amadis de Gaul, which is full of impossible adventures" (272) reinforce our sense of where the danger lies. As Harry admits, "I was prepossessed with a particular notion" (251), which kept him from seeing accurately. In a conversation with his father at the end of the work, Harry wonders why so few people manage to make important scientific discoveries in such an enlightened age. The father answers, "They may have some prejudice or favourite theory, which prevents their seeing what is before their eyes. Their failure arises from taking a wrong view of the object, or a wrong road to it" (331).

Quixotism now seems to name a strong version of subjectivity in general, and windmills serve as an emblem of both the danger and the solution. At the very center of the 1801 version, the children encounter a windmill on their way to visit their scientific uncle. No one mistakes it for a giant; indeed, it becomes the occasion for a lesson in patient observation. A nearly mechanical objectivity emerges as the antidote to quixotism. "His father looked at his watch during one minute; and Harry counted the number of revolutions, or turns, that the sails made in that time. He found, that they went round forty-five times in a minute."[81] Still, though, a book's presence hovers over this scene. Harry observes, "I feel very glad to find, that things, which I have read in that book, are like real things, and that what I have read is of use to me" (294). He is referring not to Cervantes, but to Barbauld's *Evenings at Home*, a work that the children frequently read and admiringly cite.[82] The Barbauld story in question is called "Eyes and No Eyes." In it, two boys separately walk the same route, up a hill with a windmill on the top. The self-absorbed boy sees nothing and learns nothing, while the other boy notices everything, brings back curious objects, and learns about new natural phenomena. "What a number of new ideas this afternoon's walk has afforded you," says their tutor approvingly to the externally focused, nonquixotic boy, "one man walks through the world with his eyes open, and another with them shut; and upon this difference depends all the superiority of knowledge the one acquires above the other."[83]

Edgeworth takes the opportunity to distinguish further between books that contain things that "are like real things" and those that do not. But in the 1825 sequel (in which windmills also appear), the solution is more complex than just keeping your eyes open or even choosing the right kind of books to read. It is not enough to open your Barbauld and close your Cervantes. Rather, it takes a patient, disciplined, almost mechanically precise repetition of observation and experimental practice. Also required is a suspicious attitude toward one's own pet theories. Lavish praise for Humphry Davy appears at the conclusion of this long narrative, as the father holds him up as a model for the current age: "Observe with what care and exactness this great philosopher had conducted his experiments; how cautious he was to make himself certain, by repeated trials, of the causes of the effects produced; never venturing assertions, *nor trusting to his own suppositions, till they had been verified by repeated trials*" (268, emphasis added).

Objectivity can be cultivated, then, by making oneself like a machine: the steam engine is the true hero of the 1825 sequel, repeatedly discussed and described in various settings and uses. In the sugar-house in Bristol it is "the great moving power on which all the rest depended" (247); in the area around a coal mine and an iron foundry, "steam engines seemed . . . to have the world almost to themselves. These laboured continually, in vast and various works" (165). Harry dreamily remarks, "There they are, going on all night long, working, working, working, always doing their duty, by themselves, and of themselves" (162–63).[84] Daston and Galison note scientists' preference, by the 1830s, for the machine over the human observer: the machine is "more attentive, more hard-working, more honest."[85] But Edgeworth's narrative is also strangely obsessed with the idea of interior spaces. Even machines have them. Harry not only explains to Lucy in detail the vacuum inside a piston but often returns to the air-pump, noting that "all goes back to that one great principle of the vacuum" (252). Harry's portable camera obscura, another well-known figure of interiority, also plays a role in the sequel's overarching plot. Something powerful is needed to pull the subject out of herself, to turn attention outward and to fix it on external objects for a sustained duration—for example, the operation of plot as a continuous objective disclosure.

Plot is carried out here in a highly self-conscious, self-referential way. The 1825 sequel refers often to objects and experiments from the original two-part *Harry and Lucy*. The latter had concluded with the triumphant explanation of a barometer; the sequel begins with Harry bringing home a portable barometer. Toward the end of the sequel, Lucy recalls an experiment involving condensation on a plate that the children had done with their parents in the very first part. But in *Harry and Lucy* and in other tales, Edgeworth always emphasizes the role of objects. An object can become a nexus of plot because of the characteristics it possesses in the external world: a limited perdurance over time, for instance, or portability. At the same time, the object's referential character can make it multifunctional with regard to plot. In other words, its multiple functions in the plot depend on knowing the characteristics of its referent. For example, plot continuity is foregrounded in *Rosamond*, another tale included in the original 1801 *Early Lessons* (also revised and extended by Edgeworth years later). Here, the ability of an object to function more than one time or one way in the plot seems to determine the value of its selection. The choices made by this very young girl turn out to be either prudent or imprudent, wise or unwise, largely on the basis of the qualities possessed by the objects and the various ways they can be tied repeatedly into the plot. Rosamond's opening choice of a showy purple glass jar over a pair of new shoes is unwise not only because her old shoes soon fall apart, but because the glass jar turns out to be a narrative dead end. The next time, she chooses a sewing case to hold her needles; this turns out to be a wise choice because the needle that she thereby doesn't lose becomes central to an enjoyable science experiment. It even serves to remove a splinter from her finger in a later episode. There is obviously a moral point being made about useful, functional things versus useless, aesthetic ones, but I believe that Edgeworth also deliberately conflates a certain type of usefulness in the external world with usefulness in the plot, and both of these with objectivity. Seeds and plants, for example, figure in this and other tales (as they did for Robinson

Crusoe) because they seem to be part of an independent causal nexus, one that Rosamond can assist, ponder, or examine, but not control. This is not the artificial suspense of the Gothic cliffhanger.[86] Delayed effects and hidden causes are located in objective reality—in the length of time it takes for a seed to germinate, or in the uncertainty of its happening.

Objects in these tales create social relations between people by being bought, sold, given, or shared; they can also reveal the subject's tendency to become self-absorbed to the point of isolation and even blindness. Asked why she is in tears one morning, Rosamond tries to pinpoint the day itself as the cause of her misery: "It was partly the day's fault . . . because it was so cold? [*sic*] It was the cold that first prevented me from getting up; and then my not getting up was the cause of my being in a great hurry afterwards, and of my losing my lace and my pocket-handkerchief, and of my pulling the strings of my cap into a knot, and of my being cross to Laura . . . and of all my misfortunes."[87] Though there is a certain plausibility to this account, especially if we recall certain key episodes in Austen's *Emma* or *Mansfield Park*, she proves wiser: "(pausing to reflect, she added) 'I do not think that the cold or the [cap] strings were the *real* cause of my misfortunes . . . it was my thinking that to-day would be a day of misfortunes, that made me cry the second and third time" (61). The surface lesson has to do with the need to learn self-control and self-reliance: "I think . . . it will depend entirely upon yourself, whether it is [an unfortunate day] or no" (62), her mother affirms. But the deeper lesson concerns the danger of mistaking what is merely an interior determination ("The cause is—I believe, mamma—because . . . because I think to-day will be . . . a day of misfortunes" [56]) with causality as it functions in the external world. Rosamond is learning the distinction between subjectivity and objectivity as parallel planes on which events can occur in the plot.

The type of objective nexus that Rosamond imagines here, though, is used by Edgeworth in numerous tales and in her novel *Belinda*. Such chains reveal connections, relations, causes. They also constitute a plot mode, protocol, or procedure. Often the truth emerges through small physical marks or other distinguishing features that can be noticed by a careful observer: the black edge left on a sheet of writing paper that allows an anonymous letter to be traced, or a bird's distinctive song, for example. The recognition is only a minor event, secondary in importance to the spatiotemporal conjunction of objects and their ability to bring about effects in the external world. A gardener is cheated of a prized plant to serve Lady Delacour's vanity, loses his position, and becomes a beggar; a girl who pauses on the street to give him money thus happens to meet a lady who will eventually reunite her with her estranged mother. The mother is, of course, Lady Delacour; but the gardener and his beloved cherry tree go on to contribute in other ways to the plot's denouement.

Even more important than such networks of objects are plot situations in which a person is surprised by some object or situation into revealing something within them that cannot otherwise be seen, even by themselves. A man is unusually startled when confronted with a painting of a girl. Belinda is unexpectedly dismayed when a woman's lock of hair falls out of a piece of paper in Clarence Hervey's pocket. "She examined herself with firm impartiality—she recollected the excessive pain . . . extreme anxiety she had felt . . . the species of alarm which she had felt at this discovery, opened her eyes effectually to the state of her own heart" (138–39). The novel is full of science experiments, usually performed by the children in the Percival family very much along the lines of *Early Lessons*.[88] But there are larger experiments on which the plot itself rests: for example, the experiment to see whether the dissolute Lady Delacour can indeed be reformed into a virtuous wife and mother. Experiments are protocols, procedures just as essential to the plot as the hidden truth thereby revealed (love, family relationship, identity, secret moral character). Such procedures are sometimes set up by other characters, who then watch and observe. The subject must be caught off guard, provoked into an involuntary disclosure: "If she . . . had any secret to conceal, she will betray herself on the sudden appearance of this picture" (476), one character remarks of another. This is less like a classic recognition scene than an air-pump experiment, set up and performed with witnesses.[89]

Why should such elaborate procedures or protocols that get around conscious volition be necessary? The answer has to do with the quixotism that seems endemic, even contagious in this novel. In the strangest example—one based on a real-life experiment performed by Thomas Day, friend of Richard Lovell Edgeworth—Clarence Hervey, deeply impressed by Rousseau's *Emile* and "charmed with the picture of Sophia . . . formed the romantic project of educating a wife for himself" (362). Happening on a lovely girl living in an isolated forest, he places her in a secluded house, where she is not allowed to have any contact with other people.[90] He names her Virginia, after "M. de St. Pierre's celebrated Romance" (370), and persists in viewing her as a special child of nature, despite clear evidence to the contrary. Having nothing to do all day but to read romances, Virginia herself becomes a female quixote of the classic variety (quixotism within quixotism). "Virginia devoured these romances with the greatest eagerness" (380); she dreams of knights, tournaments, and holds "exalted notions of female delicacy and honour" (385). The experiment, which he names as such, does not turn out well. After meeting Belinda, Clarence hopes that Virginia's feelings for him will turn out to be simply "a disease of the imagination, induced by her seclusion from society" (395) and the overconsumption of romances. Insufficient attachment to external things and willful overattachment to one's own private wishes or hopes (fueled by reading) are two sides of the same problem. Virginia had no "real objects to occupy her senses and understanding" (379). The same was true of Clarence until he and Belinda undertake the Lady Delacour experiment. Lady Delacour herself also shows quixotic tendencies, stubbornly viewing both herself and the objective world through what she reads (including "methodistical books" [301]). She is cured by devoting herself to investigating the truths of other people, designing the experiment that will finally elicit Virginia's true feelings and their object: a miniature portrait of a man (who is not Clarence). Quixotism and scientific experiments seem to feed on each other; yet the hope is always that quixotism will thus be overcome, will become a temporary stage rather than a permanent condition.

One cannot voluntarily display oneself because subjectivity is inherently willful, biased, overly attached to some preconceived image of the self or world. But getting a glimpse of oneself remains a goal that is achievable for some. Belinda's dispassionate analysis of her own alarm we have already seen; Clarence, too, is able to observe himself as subject and object: "He enumerated some of the various schemes he had formed in his early youth [i.e., the Virginia experiment], and humorously recounted how they had failed, or how they had been abandoned. Afterward, changing his tone from playful wit to serious philosophy, he observed the changes which these experiments had made in his own character" (275). Clarence insists, much like the father in *Harry and Lucy*, that experiments are the only way to know with certainty: "Our friend Dr X—would . . . ask me, whether it be the ultimate end of my philosophy to try experiments, or to be happy, And what answer should I make? I have none ready" (276). Though he fears "I shall pay too dear yet for some of my experiments," meaning his entanglement with Virginia, he still asserts the superiority of "those who learn from their own experience" (276) rather than from the experiments of others.[91] Clarence's most valuable experience comes in the form of self-experimentation, along the lines of Davy.

Edgeworth's plot procedures allow us to see characters as both objects and subjects. The kind of object-filled nexus that Edgeworth favors, and the use of protocols to reveal an inner truth, seem to reveal the operation of objective causal chains. An action is distinct from an event only insofar as its cause, in Jonathan Kramnick's words, "includes mental states."[92] An action is likely, though, to be an event gone wrong, an event that was brought about not by "the essential merits of the object" (255) but by some problem of distorted will or perception. Lady Anne Percival, Belinda's wise friend, notes that love "arise[s] from circumstances totally independent of our will . . . by the agreeable or useful qualities that we discover in things or in persons. . . . Or by those which our fancies discover" (240). The danger arises from jumping to a wrong conclusion based on one's own subjective priming: "From poetry or romance, young people usually form their early ideas of love . . . and *the image which they have in their own minds* . . . is cast upon the first objects they afterward behold" (255, emphasis added). Ideally, the "essential merits of the object" would be revealed via something approaching mechanical objectivity, with love as the necessary result.

Plot can reveal the essential merits and demerits of an object in this quasi-mechanical way by engineering certain kinds of procedures.[93] Subjectivity's secrets can only be elicited through the plot's maneuvering with things, circumstances, and situations.[94] This can result in a strange dilemma, however, when the plot hews too closely to one subject's desire to know. We can see this problem in William Godwin's *Caleb Williams* (1796). Godwin is clearly aware of plot as an independent principle, as a force that can generate powerful affects of suspense. This novel's plot draws its continuity largely from one character's unremitting persecution of another and from the continuity of affect thus produced: vengeful anger on one side, terror on the other.[95] But that persecution is caused by Caleb's desire to discover his employer's secrets. In his 1832 preface, Godwin gives a lengthy account of how he wrote his novel. He congratulates himself that he has achieved an "entire unity of plot" by writing it in reverse order. His goal was to raise "a very powerful interest," so he decides on a final volume showing "flight and pursuit"; he then must invent a cause for that ceaseless pursuit and decides on "a secret murder, to the investigation of which the innocent victim should be impelled by an unconquerable spirit of curiosity." He next looks for the subject of the first volume and finds it in how the murderer should have been "driven to the first act." Godwin thus gives first consideration, according to this account, to the requirements of plot, or the qualities that plot must possess in order to produce continuous, unflagging suspense.

More like Fielding than Edgeworth, however, Godwin suggests that subjectivity can be made directly visible by the representation of inner motives, by showing what characters feel, what they want or fear. He praises his own "analysis of the private and internal operations of the mind, employing my metaphysical dissecting knife in tracing and laying bare the involutions of motive, and recording the gradually accumulating impulses, which led the personages I had to describe primarily to adopt the particular way of proceeding in which they afterwards embarked."[96] There are attempts made by the plot and in the plot to provoke subjectivity into involuntary disclosure, but they do not succeed. Godwin had argued in his *Enquiry Concerning Political Justice* that thoughts were a quasi-material medium. He described where and how mental states or determinations intervene in the chain of causes that determine exactly what we do: "Thought is the medium through which operations are produced. Ideas succeed each other in our sensorium according to certain necessary laws."[97] Exactly like the mechanism of the external physical world, which "has for its medium only matter and motion," what he calls "the mechanism of man" has thought for its "real medium."[98] Godwin wants to believe that his narrative can "[record] the gradually accumulating impulses" inside the brain just as directly as it can show the operation of causal chains in the external world.

Yet his novel runs directly into the problem of objectivity because it does not set it up as a problem for the plot itself to solve. His novel relies on the first-person narrator, Caleb, to relate a growing curiosity about his employer, Falkland. Caleb notices Falkland's extreme sensitivity to certain topics, and decides to goad him to observe his symptoms and determine whether Falkland is guilty of a past crime. Obsessed with his social reputation, Falkland ultimately acknowledges to Caleb his guilt for a murder that he will never allow anyone else to know. Falkland tells Caleb so himself, and this is meant to be decisive confirmation of other behaviors that Caleb had found suspicious. Yet the story is overshadowed by the ambiguous character of Caleb, who insists on his own innocence (he is also accused of a crime, of stealing from Falkland). Godwin's novel balances subjectivity against subjectivity, Caleb's determination to find out and then to escape his employer against Falkland's equal determination to hide his crime from the world and keep Caleb under his control. This happens with the minimal intervention of objects (the important exception being the printed advertisement about Caleb that pursues him everywhere). The problem is that both men are, in a sense, quixotes, seeing in the external world only what confirms their own obsessions. The plot seems not to know what to do with objects. A mysterious chest seems to contain evidence of Falkland's guilt. But its interior is never actually disclosed by the plot. Falkland charges Caleb with stealing; evidence is found inside Caleb's trunk, but the novel never tells how it got there. Material things in the novel prove strangely insubstantial, shallow, and inconsequential. When Caleb is put in a real prison, he finds it surprisingly easy to break out of it, despite many chains and bricks. He finds it impossible to escape the invisible agency of Falkland's wrath, however, which finds him wherever he tries to hide.

The more Caleb insists on his own integrity or Falkland claims his own guilt, the less they are able to convince each other or anyone else of what they are inside. The plot cannot even seem to convince itself and collapses aporetically, as shown in the two endings Godwin writes for the novel. In the revised ending, Caleb succeeds in being believed when he proclaims his innocence at another trial: "The artless and manly story you have told has carried conviction to every hearer," Falkland acknowledges (335). In the original ending, despite telling his tale "with great artifice and appearance of consistency" (341) Caleb is placed in prison and goes mad, obsessed with "the idea . . . of mystery, of something which the understanding was incessantly anxious to penetrate, it turned it on this side and that, it tried to enter by a thousand paths, but always returned empty" (343). "Which is the man, I or my chair?" Caleb wonders: "I have dreams . . . there is one thing first, and then another thing, and there is so much of them, and it is all nothing" (346): aporia, the blocked path.[99] In these shards of thought we see Caleb haunted by the idea of an interiority that turns out to be inaccessible because it may not be there at all. Plot decays into seriality: "one thing first, and then another thing," as lifeless as Caleb himself has become. The novel neglects to establish a legitimate standard of objectivity within or through its plot and seems to suffer from that failure. While there is more than one investigation or trial scene, Godwin sees such social mechanisms of disclosure as inherently corrupt. They cannot reveal objective truth; thus, the novel is left without a means of verifying, opening up or shutting down subjectivity. However, Godwin does succeed in producing the effect of strong suspense that was his original goal: he tells us that his friend, Joseph Gerald, "had received my book late one evening, and had read through the three volumes before he closed his eyes" (354). Here suspense seems only to feed an inner drive instead of creating attachment and attention to what Davy called "minute alterations in the external world."

The problem of establishing a standard of objectivity in a first-person narrative shapes Edgeworth's late novel, *Harrington* (1817). Written to counteract the anti-Semitic stereotypes she had relied on in her earlier tales and novels,

Harrington takes up the question of surmounting ethnic or racial prejudice. As a late quixotic novel, however, it also concerns itself with the larger question of how an idiosyncratic, even monomaniacal subjectivity can be objectively disclosed by the plot. Like *Caleb Williams*, it is related by a first-person narrator, but Edgeworth's novel could not be more different. If Caleb's madness is indicated when he sees himself as "like a stone" (346), *Harrington* identifies sanity (or a postquixotic condition) with the ability to view oneself as implicated in an objective causal nexus. It rests its resolution on a distinction between the self as an object that is acted on powerfully by external objects, and an observing self, a reader of plots. The latter learns to view the former dispassionately and without distortion—a more elaborate version of the self-control that Edgeworth had tried to inculcate in her earlier tales for children. Harrington is a kind of grown-up Harry, in danger of falling inside his own obsession, or a fictional Humphry Davy lost in his own ideas.

Manipulated by a nurse-maid's "horrible tales," young Harrington is taught to feel horror at the sight of a Jewish old-clothes man. He develops a pathological terror: "A strong false association, grown into a bodily and mental antipathy which I believed to be invincible, and felt to be powerful . . . I felt actual bodily pain from my dislike of the sight and approach of the Jew."[100] Like a quixote, the child sees in the world only things that confirm his beliefs. But (as in Cervantes, whose novel appears on a table in one scene) the external world also cruelly feeds his delusions. When his mother pays the old-clothes man to go away, "good Christian beggars, dressed up and daubed" begin to appear on every side of the child: "If I went out with a servant to walk, a Jew followed me . . . I was left alone, a Jew's head was at the carriage window, at the side next me. If I moved to the other side, it was at the other side; if I pulled up the glass . . . the Jew's head was there opposite to me, fixed as in a frame" (79–80). The adult narrator understands his own psychology as having been determined by the confluence of "those small causes which early influence the imagination, and afterwards become strong habits, prejudices, and passions" (78).

The novel's plot does not entirely turn, however, on subjective events. Harrington's prejudice is eventually erased by external causes: "The slight and natural circumstances at school and the university . . . had changed my early prejudice . . . so that now I had formed what might rather appear a natural sympathy with the race of Israel" (111). But that is not the focus of the plot. When he falls in love with the daughter of a Spanish Jew, Montanero, the plot is given heft, intricacy, and duration, by obstacles to this marriage.[101] The most serious barrier has to do with the question of whether his change of heart was objectively real. Montanero points out "what is so difficult, scarcely possible, completely to conquer: an early prepossession, fostered perhaps by the opinion of many" (161). Harrington's self-certainty wobbles: "I thought I saw in Mr Montanero's and still more in the timid countenance of his daughter, a fear that I might relapse, and that *these early prepossessions which were so difficult, scarcely possible, completely to conquer* might recur. I promised myself that I should soon convince them they were mistaken" (161, original emphasis). But Harrington cannot perform actions voluntarily to demonstrate his newfound "sympathy with the race of Israel." Rather, the most significant events are those in which, as a result of his rival's secret machinations, we see dramatic relapses in public, when Harrington seems to show that he hasn't conquered them at all. Confronted, as if by accident, with images that resemble the objects of his early phobia, Harrington involuntarily recoils, crumples, or collapses: "The sight struck me with such associated feelings of horror that I started back, exclaiming with vehement gestures, 'I cannot bear it!'" (186). Persuaded to believe these symptoms to be an incurable mental illness, Montanero tells Harrington that he must be tested, or rather blind-tested: "It depends not on your will to remove the obstacle. By no talents, no efforts of yours can it be obviated; one thing, and but one, is in your power—to command your own mind. . . . We have now an opportunity of judging of the strength of your mind, the firmness of your resolution, and your power over yourself. Of these we must see proofs" (230–31). Knowing only that he is being tested for something, without knowing how, by whom, or for what purpose, Harrington manages to display calmness and self-command in a critical situation. But he cannot control his response to anything that triggers his early associations. Those responses are shown to be objective facts, genuine responses elicited by an external situation, as true as his conscious thoughts and sympathies. But they're consigned to the historical-material, bodily-mechanical Harrington, a thing that seems to exist apart from his will and his rationality, a thing that is acted on by external causes. He cannot completely conquer these early prepossessions, but he does learn how to view this object-self with a disciplined objectivity. An attitude of scientific objectivity, then, offers a kind of solution. Curiously, the object that involuntarily triggers his panic is described as "the face with terrible eyes" (217). Are the eyes terrible because they can see or cannot see? Ironically alluding to early modern beliefs in the sympathetic cure of wounds, Harrington presents the novel as "my experiments, solitary and in concert, touching fear . . . perhaps well worth noting for future use" (78).[102] But he does not undertake the experiments himself. The plot puts him inside one, or several, to be scrutinized and studied by others. It is also able to imply that its final experiment is decisive. Notably, it places Harrington in a historical incident, the Gordon Riots. In a detailed and drawn-out plot sequence, he has to reveal both self-command and the ability to trace objective causal connections in order to exonerate Montanero from a false accusation.

Between the publication of *Caleb Williams* and of *Harrington*, quixotism becomes endemic, so familiar that it can hardly be seen. But it supplies the necessary counterpart to the objectivity that the plot will offer. The protagonist of Scott's first novel, *Waverley* (1814), grows up secretly addicted to reading romances. Catherine Morland and Elizabeth Bennet are both initially guided by a quixotic "prepossession."[103] What Davy had said about the benefits of science remains true of the plot: "The relations of these [external] phaenomena are perpetually changing; and consequently they are uniformly obliging us to alter our modes of thinking" (Davy, 325). In Elizabeth's version, "People themselves alter so much, that there is something new to be observed in them for ever" (31). But it is not only someone's letter or explanation that removes "former prejudices" (378). Faithless, recidivist characters prove to be versatile and indispensable, like Rosamond's needle. They allow Austen's plots to stitch together by the end of the novel new relations of objective indebtedness as well as unity.

In a neat twist, Emma Woodhouse's quixotism lies in her infatuation with the very idea of objectivity. She is convinced that she mostly observes, with mechanical precision and accuracy, things and tendencies that are really there. We can see how she unconsciously projects her own social independence onto the external world, insisting on the operation of an independent causality. She believes that her role is, like a Newtonian deity, only to give the smallest of adjustments here and there to the self-working of natural causal forces. By the third volume, she arrives at the pinnacle of objective detachment. "Every thing was to take its natural course, however, neither impelled nor assisted."[104] This marks, of course, the height of her subjective delusion. But the novel's free indirect discourse provides a striking novelistic instance of what Daston and Galison call "blind sight."[105] Their phrase refers to the way in which inanimate mechanisms can generate images without the intervention of belief, taste, or will. Though not an inanimate mechanism, free indirect discourse likewise shows us how Emma sees without seeing or believing. Sentences that begin with variations of "Emma watched" or "Emma saw" present both what Emma thinks she sees and a truth that she cannot see because of her situation within the unfolding plot, as well as her own quixotic attachment to objectivity. Of Frank Churchill, for example, "She saw that Enscombe could not satisfy, and that Highbury, taken in its best, might reasonably please a young man who had more retirement at home than he liked" (173). The perception is correct, though the referent is wrong. Emma does not know at this point that the "best" in Highbury is Jane Fairfax. Emma's blind sight can be attributed to her conviction that she can maintain a barrier between what she wants to happen or to be the case, and the way things really and independently are. In a sense, her sight is objectively clear, but her commitment to objectivity becomes a false consciousness that makes her vision unavailable or useless.

Small touches of the plot both invoke and defeat the protocols of objectivity. In Edgeworthian fashion, Frank invites Emma to watch while he tests Jane Fairfax: "I must go and ask her whether it is an Irish fashion. Shall I?—Yes, I will . . . and you shall see how she takes it—whether she colours'" (174). But Frank stands so as to block her view of any guilty blush. Undeterred by such failures, Emma thinks she can even view herself as an object without any bias. This is how Edgeworth's Belinda had "examined herself with firm impartiality" to see whether she was in love: "The species of alarm which she had felt at this discovery [of a possible rival], opened her eyes effectually to the state of her own heart" (139).[106] Emma: "I do not find myself making any use of the word sacrifice. . . . In not one of all my clever replies, my delicate negatives, is there any allusion to making a sacrifice. I do suspect that he is not really necessary to my happiness. So much the better" (207). Combining Belinda's observational rigor with Arabella's unhinged epistemological self-confidence, Emma also examines Frank's behavior for symptoms of love in another example of blind sight: "*He* is undoubtedly very much in love—every thing denotes it—very much in love indeed!" (207). All too aware of the threat that subjective bias poses, Emma prides herself on her ability to separate her wishes, her "authority, aesthetic pleasure, or self-love" from her "objective vision."[107] That is the most dangerous, insidious form of prejudice, but she cannot observe herself truly until the plot laboriously arranges it for her. If Edgeworth had relied on artificial plot situations to reveal the hidden inner being of characters, Austen's plot finds in Harriet "exactly the something . . . required" (21). The novel assumes that Harriet's subjectivity is neither deep nor serious but a kind of material substance that is ductile on the surface but stubbornly resistant beneath. Like the surgeon's hand in Edgeworth's *Belinda*, the plot uses Harriet's words, beliefs, and actions to reveal hidden aspects of Emma. It is only as a consequence of Harriet's announcement that Emma finally becomes "acquainted with her own heart"; characteristically, what she finds there is not a wish but an axiom: "Mr. Knightley must marry no one but herself!" (320). This discovery can only be made for her by means of Harriet, who comes to figure the plot's objectivity, its indifference to subjective wishes, beliefs, or efforts. Despite temporary detours, Harriet continues on her trajectory toward the Martin state, the inertial force of the external world is confirmed, and, in a final twist, Emma's belief that she has done nothing but observe turns out to be true.

CHAPTER 8

Historical Vertigo and the Laws of Animal Motion

Edward Waverley, the protagonist of Walter Scott's first novel, grows up reading unsupervised in his uncle's large Gothic library. Just like Scott himself, whose early years provide the model for these chapters, Waverley devours histories, fictions, "the romances of chivalry, and the ponderous folios of Cyrus and Cassandra."[1] In fact, he becomes a quixote, but a secret one. Though "the reader may perhaps anticipate . . . an imitation of the romance of Cervantes," the narrator remarks, "My intention is not to follow the steps of that inimitable author in describing such total perversion of intellect as misconstrues the objects actually presented to the senses, but that more common aberration . . . which apprehends occurrences indeed in their reality, but communicates to them a tincture of its own romantic tone and colouring" (18). Waverley dreads being discovered as a quixote, but enjoys this ability to imagine himself living inside a book or a painting. His quixotism plays an important role in the plot when he falls in love with the Jacobite Flora Mac-Ivor, "precisely the character to fascinate a youth of romantic imagination" (116). Even more consequentially, it shapes Waverley's response when he meets "a prince, whose form and manners, as well as the spirit which he displayed . . . answered his ideas of a hero of romance" (193). But the prince and the enterprise are both historical facts. Waverley kneels to an actual figure, Charles Edward Stuart, who returned to Scotland in 1745 to try to reclaim the British throne. The subsequent battles that occurred at Preston, Clifton, Culloden, all described or alluded to in the novel, actually happened. But why place a quixote in the pages of history? Or frame a quixotic plot around a historical event?

An obvious answer would lie in the quixotic nature of the Jacobite enterprise itself, using the term loosely and in the modern sense. Even Waverley cannot help noticing its implausibility; as the army lines up outside Edinburgh, he feels "astonished at the daring attempt of a body not then exceeding four thousand men . . . to change the fate, and alter the dynasty, of the British kingdoms" (215). But the more extensive answer that this chapter will pursue has to do with several factors that I discussed in the previous chapter. The first concerns the way that novels come to normalize quixotism in their plots as a "common aberration," posing subjectivity against objective truth. The above account of Waverley emphasizes the many degrees of subjectively distorted apprehension that can take place well before someone reaches the level of a Don Quixote.[2] The other factor has to do with the changing sense toward the end of the eighteenth century of what counts as knowledge: that new attention to the invisible forces and conditions that shape an organism's development and to which its coherence responds. Taken together, these point the way toward the historical novel. In Scott's hands, the historical novel will examine the way that your sense of what's happening around you will be subjectively distorted. It will give equal emphasis to the deep, barely visible forces that make the environment, or the current conditions of action, less a stable ground than a continually shifting, unsettled one. Together, the workings of subjective distortion and the movement of objective historical forces create a rich novelistic dialectic, shaping and struggling against each other at multiple levels.

Waverley (1814) will play an important role in this chapter's argument. There are good reasons that Scott's oeuvre came to be known as "the Waverley novels," and Scott himself as "the author of the Waverley novels," for this first novel provides in many ways a model for what followed.[3] In its final chapter, "A Postscript, which should have been a Preface," we can find an analogy fundamental to this new form. Scott compares the previous half century's social and political changes in Scotland not to a revolution, rupture, or upheaval, but rather to an experience of floating down a fast-moving river: "Like those who drift down the stream of a deep and smooth river, we are not aware of the progress we have made until we fix our eye on the now-distant point from which we set out" (340). He continues, "Such of the present generation as can recollect the last twenty or twenty-five years of the eighteenth century, will be fully sensible of the truth of this statement." At the risk of making too much of this simile, I want to suggest that through it we can find a model for what plot can now do in the novel. With the continual backdrop of the master-plot of extinction, plot, like natural history, tries to preserve specimens—of Jacobites, for example.[4] Scott avows his "purpose of preserving some idea of the ancient manners of which I have witnessed the almost total extinction."[5] It thus needs to posit a "total world within the entire circumference of which [an action] moves," as Hegel put it.[6] In other words, the novel cannot simply display these characters alongside each other to be compared in terms of static identities and differences; the plot must show the conditions to which they are responding. It must also show how those conditions never remain the same.

The plot of the historical novel also tries to mark the distance between then and now, and measures the rate, extent, and forms of change.[7] But rather than assume, like Austen's novels, that the rate and features of change are determined by the substance involved, the historical novel's plot looks for deeply buried forces that suddenly become manifest in certain conjunctures, driving that change and testing human subjects. Like the experiments artificially set up in Edgeworth's plots, or the agonizing surveillance we find in Burney's, it reveals the inner substance of characters. The process is involuntary and the results surprising even to themselves. In the *Heart of Midlothian* (1818), for example, Jeanie Deans, a young woman in rural Scotland, walks to London to beg a royal pardon for her sister.[8] In Georg Lukács's words, she then "returns to everyday life, and never again does she experience a similar upsurge in her life to betray the presence of such strengths."[9] People cannot display their hidden capacities through an act of will. Even orchestrated fictive events that occur at the level of everyday life would not be adequate. Only a complex, objective historical and fictional conjuncture can activate them. At the start of *Old Mortality* (1816), the fictional hero Henry Morton lives oppressed by his miserly uncle and even the housekeeper's petty tyranny: "The current of his soul was frozen by a sense of dependence, of poverty."[10] Morton makes an impulsive decision out of a complex web of private and public feelings; it leads him to assume a role in the Covenanters' army in a turbulent, post-Restoration Scotland. The change is called forth both by his buried past, his lineage, and by an intricate historical situation, in which the Covenanters have been declared insurgent rebels by a newly restored monarchy distant from Scotland: "That moment made a singular and instantaneous revolution in his character. . . . His character was as effectually changed as the appearance of a villa, which, from being the abode of domestic quiet and happiness, is, by the sudden intrusion of an armed force, converted into a formidable post of defence" (192–93). Fiction stops the historical clock at a particular moment. That conjuncture then does the work of intricate fictional protocols: it reveals the hidden depth of characters.

Perhaps most importantly, however, the plots in these novels move characters around. They focus on displacement. This is how Scott describes the plot of *Waverley* in his General Preface, added in 1829: "The tale of Waverley was put together with so little care, that I cannot boast of having sketched any distinct plan [i.e., plot] of the work. The whole adventures of Waverley, in his movements up and down the country . . . are managed without much skill. It suited best, however, the road I wanted to travel" (354). In a striking contrast with Godwin's method of composing plots carefully from back to front to maximize suspense, Scott assimilates plot to simple movement. This may look casual, even trivial, as a way of describing plot, until we recall that movement provides something more than a simile for history. It also offers a way to link history with the genre of the novel, as Scott himself defined the latter. In his long 1815 review of Austen's novels, Scott traces how this new genre emerged from romance in two stages. In the first stage, "Earlier novels differed from those now in fashion, and were more nearly assimilated to the old romances. . . . By the studied involution and extrication of the story, by the combination of incidents new, striking and wonderful beyond the course of ordinary life, the former authors opened that obvious and strong sense of interest which arises from curiosity."[11] In "the last fifteen or twenty years," however, "a style of novel has arisen . . . neither alarming our credulity nor amusing our imagination by wild variety of incident." The plots of Austen's novels rely only on "general experience," "common occurrences as may have fallen under the observation of most folks," and "that which is daily taking place around" the reader.[12] Note the familiar placement of the historical break or transition around the end of the eighteenth century.

As the novel now takes up the minutiae of quotidian experience, movement enters the picture or the plot differently. It is not the case that laws of motion cease to matter. The plot still has to affirm, in the words of Newton's popularizer Willem s'Gravesande, that "motions are performed by certain rules, and always subject to the same laws. . . . Even in those things which appear to us wholly fortuitous and uncertain, certain rules are without doubt observed."[13] But the everyday experience of movement, for instance, when drifting down a river, is very different from the way that natural philosophers thought of motion. Even in his time, however, Newton's popularizers had to reckon with the fact that to us, it feels and looks as though the earth is standing still. Perception and cognition are frequently at odds. Benjamin Martin had written that "we may not only have an idea of motion in bodies when they do really move, but also when they *do not move*, but are absolutely at rest. This ambiguity in our ideas of motion renders us subject to many and very great mistakes in our common notions or judgment of things."[14] He uses the example of two ships side by side: when the ship that we're on moves forward, we project our own movement onto the ship that is actually at rest and see it moving backward. This quixotic projection is now known as vection; the more usual example given is that of two trains side by side, only one of which is moving.[15] Later in the eighteenth century, the natural phenomenon of vection was assimilated to that of vertigo and studied by physiologists. Vertigo or vection provides a familiar, everyday example of subjective bias or distortion, a kind of quixotism of the body. We can't tell if we're moving or the environment around us is moving. We can't tell whether we're moving forward or backward. Vection relies on the interplay of habit and environment, as we'll see, and thus links the novelistic representation of everyday experience with the idea of history as movement.

The plots of Scott's historical novels can focus, then, on the unconscious or half-comprehended experience of movement: not only spatial in nature but social, political, economic, and cultural movement. History comes to be conceived not as a succession of static periods or stages but as perpetual uneven transition, as the shifting of the ground under one's feet, imperfectly registered, incommensurable with individual perception, volition, or cognition.[16] History in the plots of these novels is unsettled, full of hidden, latent forces underneath where you stand: this is why the Jacobite rebellion, what might seem to be a minor challenge to established authority, becomes paradigmatic for the Waverley novels. At the same time, in its immutable sets of recorded dates, names, and events, history lends a kind of objectivity to the novel's plot that legitimates it, constrains it, and establishes a productive dialectic with subjectivity.[17]

Movement or motion provided, in at least one case, a curious means of dismantling subjective experience, of reducing it to rates and patterns of vibration. Erasmus Darwin, friend of Edgeworth's father, and grandfather of Charles Darwin, is often discussed by current scholars for his long poem, *The Botanic Garden* (1791), which communicated the insights of botany, geology, mineralogy, and even the history of the steam engine in poetic form.[18] But Darwin's extensive scientific undertakings remind us of the relative ease with which scientific boundaries could still be crossed at the end of the century. Darwin had studied with William Cullen in Edinburgh, an important center for medicine and science, and a world to which Scott was well-connected.[19] Although Darwin was not quite of the same stature as someone like Davy, his authority as a poet and natural philosopher was widely acknowledged. Four years after *The Botanic Garden* and at the height of his success, Darwin published *Zoonomia; or, the Laws of Organic Life* (1795–96). In this work, Darwin aims to lay out the unique laws that govern "animal motions," or the motions of living beings. Other sciences, he notes, mistakenly tried "to explain the laws of life by those of mechanism and chemistry; they considered the body as an hydraulic machine, and the fluids as passing through a series of chemical changes, forgetting that animation was its essential characteristic."[20] He offers *Zoonomia* as an all-encompassing "theory . . . that should bind together the scattered facts of medical knowledge" (1–2). Here, I'd like to examine *Zoonomia* as a natural philosophical partner for the novel as Scott develops it. As we'll see, Darwin uses the phenomenon of motion to account for all the features of organic life and subjective experience. Movements that happen within the body explain all our perceptions and sensations as well as the functions that keep us alive. Darwin gives particular attention, however, to the experience of moving the body through space. It turns out that standing upright and moving through space both depend on a history that transpires at intimate and at public levels.

Darwin situates his project in relation to existing branches of natural philosophy in order to claim to be doing something entirely new. He starts with a familiar distinction between matter and spirit, and then between "primary and secondary . . . motions of matter": "The secondary motions are those, which are given to or received from other matter in motion. Their laws have been successfully investigated by philosophers in their treatises on mechanic powers. . . . The primary motions of matter may be divided into three classes, those belonging to gravitation, to chemistry, and to life: and each class has its peculiar laws" (5). The laws of the first class were figured out by Newton; chemistry's laws have yet to be explained; the laws of the third, "all the motions of the animal and vegetable world," (6) are his subject. This is a "yet unconquered field of science," he claims; "animal motions . . . have no mechanical proportion to their cause," nor can they be included within the phenomena of gravity or chemical reactions (14–15). His aim is to show how "the great variety of animal motions can be duly arranged into natural classes and orders" (14). The language of representation, ordering, and arrangement suggests that he's still working within the classical episteme; but instead of looking at the variety of visible animal shapes, forms, or parts, he aims to uncover how the parts of all living things move internally and externally, in response to surrounding conditions.

In the section on the "laws of animal causation," Darwin explains the three kinds of internal motions we experience, all of which arise from the contraction of "fibres, which constitute the muscles and organs of sense" (30). If something external stimulates a sense organ, this produces the kind of contraction called "irritation"; in response to that original contraction, depending on its quantity, we feel a secondary motion of pain or pleasure: "These constitute sensation." Finally, "A certain quantity of sensation produces desire or aversion; these constitute volition" (31). He does distinguish between motions of the muscle or nerve fibers and "motions of the sensorium," which are not mechanical, nor "vibrations . . . nor condensations," but "changes or motions . . . peculiar to life" (33). Darwin then explains how these motions are bound or related to each other. "When fibrous contractions succeed other fibrous contractions, the connexion is termed *association*; when fibrous contractions succeed sensorial motions, the connexion is termed *causation*; when fibrous and sensorial motions reciprocally introduce each other in progressive trains or tribes, it is termed *catenation* of animal motions. All these connexions are . . . produced by habit; that is, by frequent repetition" (12–13, original emphasis). Where earlier natural philosophers had hesitated before that tricky gap between mind and muscle, Darwin does not. He asserts that a movement of the mind or sensorium can be a real cause of a movement in the body. It is interesting, however, that Darwin invents a category called "catenation" and distinguishes it from both causation and association. Catenation refers to a kind of extended feedback loop in which an idea of something or a sensation of pain, for example, leads to a physical motion, which then produces another idea or sensation of pain (or pleasure), and so on, without any specified limit. These catenations possess several important features. They are produced by repetition: "All the fibrous motions, whether muscular or sensorial, which are frequently brought into action together, either in combined tribes, or in successive trains, become so connected by habit, that when one of them is reproduced the others have a tendency to succeed or accompany it" (49). They exist at a deeper, more essential level than the material of the body or the contents of the mind and thus form the basis of personal identity: "Our identity is known by our acquired habits or catenated trains of ideas and muscular motions . . . in those alone can our identity be supposed to exist . . . every deduction of reasoning, every sentiment or passion, with every fibre of the corporeal part of our system, has been subject almost to annual mutation; while some catenations alone of our ideas and muscular actions have continued in part unchanged. . . . It is these habits of action, these catenations of ideas and muscular motions, which begin with life, and only terminate with it" (134). Catenation is character; character is nothing but this plan or pattern. At the same time, these catenations occur beyond the grasp of consciousness. Far from being in our control, catenation refers to the kinds of extended sequences that give rise to our desires and volitions. Is it possible to link catenation with plot as well?

Darwin's work in some ways anticipates what will be seen much later as the novel's unique predicament: the need to bridge in its plots both "microsensory events," "the infinite multiplicity of virtual connections," and the "causal concatenation of events" on a larger scale.[21] This is how Jacques Rancière describes what modern novels try to do. This project, he argues, emerges only along with a fundamental democratization of both society and representation: the possibility that everyone and anyone, no matter how unimportant, could have a plot. Darwin's work, like Scott's historical novels, participates in this work of opening up plot and embedding it more intimately in the texture of ordinary experience. What Scott identifies as the modern novel (i.e., Austen's novels) eschews "wild variety of incident" or "minds, manners, and sentiments, greatly above our own." Thus, Scott suggests, the novelist now "encounters peculiar difficulty" in trying to concoct a "striking and interesting" plot out of the minutiae of daily life.[22] Darwin's discussions both draw on and aim to explain the same kinds of everyday experience that the novel represents, the phenomena "which occur in our daily exertions" (31). For example, *Zoonomia* includes multiple experiments that the reader can perform anywhere, using their own body (looking at an image printed on the page and then closing the eye, or pressing the eyeball in the dark). Most relevant for the novel's purposes, however, *Zoonomia* presents the body as linked by habit even in its smallest motions to a specific set of circumstances: as specific as the arrangement of furniture in your room, which you automatically know how to avoid bumping into. In this sense, catenations not only connect our ideas and muscular movements but also connect our ways of moving and acting to a larger external configuration that we inhabit. And that configuration is subject to change. Indeed, it becomes axiomatic in Scott's novels that this configuration is in the process of changing.

The discussion of vertigo is perhaps richest in possible connections with a novel's plot. Darwin's idea of vertigo, which can be read in the context of quixotism, conceives of this phenomenon as more than a condition of temporary dizziness or nausea. Vertigo encompasses the entire problem of orientation, the body's standing in the world. It unveils the strangely blind, tenuous, lagging relation that perception bears to that world. Vertigo shows how moving yourself through space relies on the illusion of stasis, the illusion that the world is physically stable. Yet this illusion is notably fragile; it's all too easy to lose our grasp of it. Standing upright is just as tricky, it turns out, as moving. Darwin begins by stating that we need to consult the objects around us to determine whether we're standing upright or not. Usually we do this through vision. Proprioception, or the internal feelings of the muscles, is a much "less accurate" source of balance; visual reference to objects is more accurate, Darwin argues (232). We can confirm this when we remove visual cues, surround ourselves with confusing ones, or stand on top of a high tower and feel dizzy. But when we try to balance, or to walk without staggering, what we actually take in with our eyes is not the external world itself, but the "apparent motion of the objects within our distinct sphere of vision" (232). A blindfolded person can't tell whether he's standing up straight: "He inclines so greatly, before he is warned of his want of perpendicularity by the sense of touch, not having the apparent motion of ambient objects to measure this inclination by" (232) that he will find it impossible either to stand up straight or to walk in a straight line. Darwin carefully uses the phrase, the "*apparent* motion of ambient objects," because this is exactly the point at which vertigo begins. Vertigo is represented not purely as an internal ailment, as something going wrong inside the body, but as the consequence of a disruption in the context of one's own standing and moving. It raises questions of time lag and aftereffects. It is the register of change, the physiological afterimage of displacement. It also registers the degree to which we need to learn and internalize a certain kind of fictionality. Vertigo happens, basically, when we can't hang onto that particular fiction, when that fiction collapses. It arises when we can't help noticing that both the ambient movement and the ambient stasis we see are largely fictional, that what we see isn't real.

What enables you to stand upright or move without falling is an ability to train yourself *not* to see what your eyes actually see. When we move ourselves through the world, we see "the jumping and progressive motions" of things around us even though these things are standing still (234). These are the "apparent motions of ambient objects." We need from "experience to learn the peculiar mode of motion of any moving objects," and, even more importantly, we have to learn "to distinguish our own real movements from the apparent motion of objects. Our daily practice of walking and riding . . . soon instructs us with accuracy to discern these modes of motion, *and to ascribe the apparent motions of ambient objects to ourselves*; but those, which we have not acquired by repeated habit, continue to confound us" (233, emphasis added). We have to learn, in other words, to stop projecting our own motion onto the world. Another way of putting this is that we have to learn to project onto the external world an objective stasis or rootedness that we do not actually see. In this sense, everyone has to be a quixote in order to stand or walk without falling down, to see the world as stable.

This capacity, Darwin notes, is easily lost in an unfamiliar setting or context: "When first an European mounts an elephant . . . whose mode of motion he is not accustomed to, the objects seem to undulate, as he passes, and he frequently becomes vertiginous" (234). It can also be acquired, though Darwin himself seems to experience a kind of cultural vertigo when he notes that "the dervises [*sic*] in Turkey" are "habituated to this kind of [rotating] motion" and can perform "swift gyrations" without feeling dizzy (235). It is also lost "when both ourselves and the circumjacent objects are in real motion" (234). When crossing over a river, or going on a ship, "where the movements of ourselves, and the movements of the large waves are both new to us, the vertigo is almost unavoidable" (234). These examples also suggest an intimate relation between vertigo and empire. The phenomenon that Darwin refers to here, of losing the ability to tell whether you're moving or the world is moving, is the type of vection that Scott uses in *Waverley* to describe the experience of historical transition. The plot of many Waverley novels might be summarized in this sentence from Darwin: "When we are surrounded with unusual motions, we lose our perpendicularity" (234). These plots concern themselves with uprightness as a way of being in the world and with the adjustments needed to maintain that posture as the ground shifts.

In other ways, too, Darwin's discussion of vertigo intersects with the concerns of plot. He notes as one of the "peculiar circumstances attending this effects of moving objects" the fact that vertigo can be precluded by a certain kind of concentrated, conscious focus: "In an open boat passing from Leith to Kinghorn in Scotland . . . I observed, that the undulation of the ship, and the instability of all visible objects, inclined me strongly to be sick . . . but as often as I bent my attention with energy on the management and mechanism of the ropes and sails, the sickness ceased" (235).[23] A cure is found in attending to this conjuncture of action and structure, "the management and mechanism of the ropes and sails": what the sailors are doing or trying to do, but also how the various parts are connected together, how pulling on one thing affects another. Plot can document and even produce the condition of vertigo but also, it seems, provide a cure, if we attend to it as a concatenation of moving parts, a form or mechanism. In this may lie, in fact, one source of its powerful appeal.

Unlike the sublime, which appeals to either the bodily or transcendental structures behind aesthetic experience, vertigo testifies to an innate, embodied historicity. It seems to be linked more closely to questions of knowledge and praxis than those of aesthetic experience or judgment. Its immediate physiological cause Darwin locates in the phenomenon he calls "spectra": "a continuation of the motions of the optic nerve excited by those [unusually moving] objects" (237).[24] Long after the object that we saw (or heard) is gone, its image continues to reverberate, with actual motions in what he calls the sensorium: "All the other classes of motions continue to go on, some time after they are excited" (236). The longer you saw something, the longer you'll continue to see it afterward. Darwin cites the example of a person revolving on one foot: "The longer he has continued to revolve, the longer will continue these successive motions of the parts of the optic nerve" (237). But is the spectral afterimage or motion subjective or objective? The motion is objective; the image is subjective. This is more of a problem, Darwin thinks, in moments of quiet or of the suspension of action and movement, "at these intervals of reverie, or on the approach of sleep" (241). It is then that people may "attend to the apparent motions, and to the battement of sounds of the bodies around them, and for a moment mistake them for those real motions of the ship, and noise of wheels" (240–41).[25] Perceptions and sensations are not simply subjective: they possess their own independent temporality or lifespan. They are rooted in history, and offer a foundation on which the historical novel's plot builds its realism.

The kind of habitual concatenation of eye and muscle that allows you to walk through a room without bumping into things is not the same thing as joining a rebellion against an established government. But Darwin's discussion allows such actions to be connected, grounding them in the texture and internal catenations of the body. He is not particularly interested, like earlier natural philosophers, in correcting misapprehensions. In 1754, one had noted, "If the eye be in motion, an object before it, tho' at rest, will have its image move over the retina, and so produce a deceptive idea of motion. . . . Unless we are well appriz'd of the true state and circumstances of bodies which appear to move, or not to move, both in respect of them and our selves, it is impossible the judgment we form of their motion should be correct, but very precarious, and mostly full of illusion and error."[26] Rather, Darwin wants to account for how we handle the tricky relations between perceiving external motions, sensing internal motions, and just moving or standing without falling down—how we manage to build these precarious judgments into our normal lives, and what happens what we fail. Darwin does not address whether "catenations of ideas and muscular motions" (134) might be shared or heritable. It is here that Scott picks up the thread.[27]

We can read Darwin's *Zoonomia* as a kind of theory that applies to Scott's historical novel, as well as an example of the ways in which natural philosophers were thinking about bodies in motion at the end of the eighteenth century. Darwin's work examines why it's hard to stand upright, and why it's difficult to tell whether we ourselves are moving, or the movement is happening outside us, by itself. The reason has to do with more than the organic structure of the nerves and muscles, as Darwin theorizes them. There is the working of habit: habit creates and hardens those catenations, as we have seen: sequences of actions and feelings leading to more actions and feelings—a kind of plot that's embedded within the body itself. But there is more than that. There is also a phenomenology of uprightness. We judge our own position (upright, leaning, falling over) and our movements by what we see happening immediately around us: other things, bodies, and structures. We constantly check, calibrate our postures and movements with reference to this external world. The problem now is that this external world can appear to move when it does not; we've come full circle from the problem that Newton's popularizers had to explain.

In the plot of *Waverley*, as mentioned above, Scott seems to replace action with motion. Where we expect to find choice and decision, we find at best impulse and post hoc rationalization. The protagonist's name indicates a lack of groundedness or stability. As one character says to him, "I cannot tell what to make of you; you are blown about with every wind of doctrine" (237). His mobility is not just ideological, however; it is physical and strikingly passive, as Alexander Welsh pointed out in his study of the Waverley novels.[28] While Welsh finds a political significance in this passivity, I want to suggest that it plays a different role. For most of the plot, Waverley is being carried along. Sometimes this happens metaphorically, as at this ball at the height of the Pretender's success: "He was supported and excited by kindred spirits, who felt the same impulse of mood and time . . . hurried along by the torrent" (209). In battle, it happens more literally, as he's carried forward within the phalanx formed by the clan he's adopted: "Waverley felt his heart at that moment throb as it would have burst from his bosom. It was not fear, it was not ardour—it was a compound of both, a new and deeply energetic impulse, that with its first emotion chilled and astounded, then fevered and maddened his mind . . . the clans rushed forward, each in its own dark column" (225). But most often, he's literally dragged and carried around by other people: sidelined with a sprained ankle, then taken prisoner by the English, then kidnapped by the Jacobites (and again injured). He often does not know where he is, or where he's going, in what direction or how fast. The movements occur outside Waverley's volition and sometimes beyond his consciousness: "The velocity, and indeed violence, with which Waverley was hurried along, nearly deprived him of sensation" (177), in a typical moment. His sensations cannot keep up; he cannot stand up metaphorically because he does not recognize what's happening around him.

Waverley's vertigo provides the basis of the plot's deep realism. We can make several connections back to Darwin's theory. Waverley cannot tell whether he's moving, or the world around him is moving, and the narrative makes much of this uncertainty. This is how Waverley ends up officially joining the Jacobites (for a while): "Rejected, slandered, and threatened upon the one side, he was irresistibly attracted to the cause which the prejudices of education, and the political principles of his family, had already recommended as the most just. These thoughts rushed through his mind like a torrent, sweeping before them every consideration of an opposite tendency . . . and Waverley, kneeling to Charles Edward, devoted his heart and sword to the vindication of his rights!" (193). This moment appears to be less of a conscious choice or decision than a catenation, as Darwin describes it, when a whole chain of feelings and actions happen all together. The catenation is inside Waverley; it is also outside him. The plot is an elaborate catenation of movements. For much of the novel, we don't know who is actually behind Waverley's various kidnappings, rescues, and forced movements. At the end, we learn that Rose, his future wife, concerned for his welfare, arranged by letter for these to happen. A host of intermediary figures is involved, including Charles Edward, who "caused the most positive orders to be transmitted to Donald Bean Lean [an independent 'freebooter'], to transmit Waverley, safe and uninjured." Donald "durst not disobey . . . transmitted orders to his lieutenant to convey Edward to Doune. . . . The governor of Doune was directed to send him to Edinburgh as a prisoner" (307–8). We see a chain of feelings and motions.[29] In the last instance, however, historical forces move Waverley around, or make him feel as if he were moving.

The plot sets Waverley's movements in counterpoint against much larger and more complex motions in the world that surrounds him; the resulting vertigo becomes another source on which the plot's realism can draw. In addition to the constantly shifting political alliances in London that form a distant backdrop to the action, we witness numerous scenes when large numbers of bodies assemble, swirl, charge, disperse. Objectively, something is happening, but here we share the vertigo, as we can't tell which things are moving or in what direction. Early in the novel, Waverley is taken to a hunting party; we see "various parties of Highlanders, climbing rocks, struggling through copses, wading brooks, and traversing thickets" to round up the deer; the deer form "a most formidable phalanx" and then charge at the hunters, forming a "tide . . . absolutely irresistible." The next morning, Waverley observes the party dissolve: "The various tribes assembled. . . . Some . . . were seen winding up the hills, or descending the passes which led to the scene of action. . . . Others made still a moving picture upon the narrow plain, forming various changeful groups, their feathers and loose plaids waving in the morning breeze" (117–19). It is later revealed that this hunting party was an excuse for Jacobite leaders to assemble and make plans for the rebellion. But beyond this, the description anticipates and captures in a single sequence the actual history of the Jacobite rebellion as a series of movements. Together with the chapter that describes the assembling of the Jacobite army before the battle at Preston, it illustrates perfectly the kind of situation that typically creates vertigo, in Darwin's model. More than pageantry, it displays historical vection. Nothing around Waverley is standing still; everything is in motion. The picture itself is moving, changing, each of its components shifting around. What does it mean, then, to stand upright? How is it possible?

The quixotism of this first protagonist, his "wild visions of chivalry and imaginary loyalty" (163), flows seamlessly into a larger model of historical consciousness as a key element of plot. Those especially who have been shaped by turbulent transitions, who have lived through struggles over religion, dynasties, or forms of law, can only perceive whatever happens through the "tincture" of their past experiences. They continue to feel the movements of history long after they have actually ceased. Scott's characters are not only stamped and often disfigured by history: they try to resist the next round of change or anticipate it, embrace or escape it—but with a certain inescapable blindness. The plots arise from this disjunction between the movement of history and the way that it's lived or experienced. The novels show characters in the act of gaming out what's going to happen, often in the very moment of struggle or conflict. They try to calculate political and social trajectories, future displacements, resurgences. If they turn out to be correct, it has to do more with luck than with rationality. Others fail to apprehend the transitions occurring around them, even as they're swept along in it. Everywhere, the plots show us characters mentally stuck in the past who are nevertheless being pushed along; or others who think they're pushing forward when they're actually standing still or moving backward.

The large-scale movements that comprise history are sometimes described from a third-person point of view. A compelling instance can be found at the beginning of *The Heart of Midlothian*, which describes the historical Porteous riots. A multitude has gathered in the center of Edinburgh to witness the hanging of Captain Porteous, the reviled captain of the city guard: "The area of the Grassmarket resembled a huge dark lake or sea of human heads, in the centre of which arose the fatal tree."[30] But the condemned man does not appear, as he has been granted a last-minute reprieve. "The crowded populace, as if their motions had corresponded with the unsettled state of their minds, fluctuated to and fro without any visible cause of impulse, like the agitation of waters, called by sailors the ground-swell. The news, which the magistrates had almost hesitated to communicate to them, were at length announced, and spread among the spectators with a rapidity like lightning" (35).[31] The description of the crowd as "[fluctuating] to and fro without any visible cause of impulse" conflates mental with physical movement and portrays the cause of both as immanent. The crowd unites and becomes one entity, even without acts of communication among themselves. At the news of the reprieve, "The assembled spectators of all degrees . . . uttered a groan, or rather a roar of indignation" (35).[32] They even think as one, and the narrative's free indirect discourse ventriloquizes their response to the news: "Is this to be borne?—would our fathers have borne it?" (36). The crowd is both one and many: there are "stifled mutterings, which each group maintained among themselves, and which were blended into one deep and hoarse murmur which floated above the assembly" (36). We somehow hear both the muttering of a small subgroup and the murmur of the entire group at the same time. Scott wants to maintain as many levels, scales, or kinds of agency or aggregation as he can. There are well-marked social ranks within this crowd: from "the more decent class of citizens," property owners, all the way down to "the rabble." But there are also groups that are numerically constituted, as well as individual particles whose behavior is consequential: "It was noticed at the time . . . that, while the mob were in the act of dispersing, several individuals were seen busily passing from one place and one group of people to another, remaining long with none" (37). These "individuals" have no class markers; for now, they are simply single units of the crowd, circulating rapidly. Nothing seems to happen despite this ominous behavior: "If . . . it was the intention of these men to stir the multitude to any sudden act of mutiny, it seemed for the time to be fruitless. The rabble, as well as the more decent part of the assembly, dispersed, and went home peaceably" (37). *For the time*: this is the key phrase, though easily overlooked. For they will return that night, break into the prison where Porteous is being held, and perform the execution themselves.

In the motions so carefully described, the Grassmarket scene does not so much relate a singular event as reveal the necessary conditions of plot in the historical novel. Note that national and class characteristics can be momentarily submerged within physical events such as aggregation, circulation, dispersal. Change occurs in history because social identities and arrangements are also less settled than might appear to a casual observer. While many are set in their ways, lagging in their perception of the present, the motions of minds and bodies can change, sometimes abruptly, at the individual and collective level. At this scale of optical resolution, as it were, we see the movement of units and of masses. The crowd sways back and forth, it breaks apart into smaller groups, and at the end, individuals circulate rapidly. These microevents are linked to others in the plot as cause and effect. We could say that history is simultaneously figured *as* movement and resolved *into* movement or movement-sequences. This is how motion becomes more than simply a figure or an analogy for history. It seems to occur at the level of history's elemental substance. This kind of fluctuating, swaying movement is also literal. It is held out, in other words, both as an event in the plot and as the physical, material mechanism through which historical change occurs.

Such scenes contain in a way the essence of the historical novel. Rather than simply inserting fictional characters in a historical event, Scott creates a dialectic between the objective movements of history and the motions and subjective apprehensions of individual characters caught up in them. The primary challenge that the novel faces when it takes historical events as its subject matter (not only events of political history but extinction-events, as we have seen) is not only the creation of plot-continuity, but the creation of openness within a set of facts that already exists. In the simplest terms, that means restoring suspense to the recounting of something whose sequence and outcome are already known. It becomes axiomatic for Scott's novels that history is always in an *unsettled* state. The ambient movement you thought was projected may be real. The discovery of this can be gradual, or it can be itself a startling event: you thought everything was settled long ago, for example, but it turns out not to be the case. Waverley is shocked when he learns that a rebellion against the long-established Hanoverian government is in progress. Because Scott's novels posit this continual flux and fermentation, the question of causality is often submerged within the event of movement. In other words, momentum enters the plot in its own right as a powerful phenomenon. In a favorite analogy, Scott compares the course of events with "the progress of a stone rolled down hill": "It moves at first slowly . . . but when it has attained its full impulse . . . it smokes and thunders down . . . becoming most furiously rapid in its course when it is nearest to being consigned to rest" (*Waverley*, 331). In this instance, Scott uses it ironically to describe the accelerating pace of his own narration as he nears the end of his novel; but he uses it in other places to account for the way that events seem to have their own momentum, to provide a sort of decoy or even replacement for causality.

Not all of Scott's plots are devoid of action; not all his heroes are as passive as Waverley. But even for those who take action, the paradigm of action becomes that of simply setting something into motion with a small push, or, more notably, of pausing or stopping something that has already gathered momentum. In *The Bride of Lammermoor* (1819), history is represented in terms of inertial movement, and action consists of attempts to stop it, accelerate it, or lose oneself in it. The novel is set in Scotland shortly after its 1707 union with England, when (as always) arrangements are unsettled and uncertain. The opportunistic newly made Lord Keeper, Sir William Ashton, has managed somehow to acquire nearly all the lands of the Ravenswoods, the traditional rulers of the area. It is not clear how that has happened. The young Master of Ravenswood says to Ashton, "How they have glided from us by a train of proceedings that seem to be neither sale, nor mortgage, nor adjudication for debt, but a nondescript and entangled mixture of all these rights . . . until our interest in our hereditary property has melted away like an icicle in thaw—all this you understand better than I do."[33] But Ashton himself seems to be more the beneficiary of this process than its instigator; the novel notes that the Ravenswoods' fortune "became greatly declined from its splendour about the middle of the 17th century" (26), even before Ashton came on the scene. Decline can always be traced back even earlier. The novel's plot suggests that movement can be paused, deflected, or accelerated under certain conditions; but volition plays a very limited role.

Here is how Ravenswood meets Ashton's daughter, to whom he will rapidly and inexplicably, even to himself, become engaged; that engagement, forcefully opposed by Ashton's wife, leads directly to the novel's tragic conclusion. Ashton and Lucy are out walking when a wild bull (the history of the species in that location is carefully noted) charges toward them. Ravenswood happens to be close by, though unseen, and kills it with a single precise shot:

> It seemed inevitable that the father or daughter, or both, should have fallen victims to the impending danger, when a shot from the neighbouring thicket arrested the progress of the animal . . . [and] proved instantly fatal. Stumbling forward . . . the progressive force of his previous motion, rather than any operation of his limbs, carried him up to within three yards of the astonished Lord Keeper, where he rolled on the ground. . . . Her father was almost equally stupified, so rapid and unexpected had been the transition . . . which did not permit him distinctly to understand what had taken place. (56)

It is never explained why Ravenswood does this; he only needed to do nothing, and his sworn enemies would have been conveniently eliminated, as Lucy herself reflects afterward. No one understands what just happened, "so rapid and unexpected had been the transition." The bull's motion seems to carry its own cause within itself; once started, a motion continues, and even accelerates. Ravenswood's character embodies the same principle: "His was a mind unwillingly roused from contemplative inactivity, but which, when once put into motion, acquired a spirit of forcible and violent progression. Neither was his eagerness proportioned in all cases to the motive of impulse, but might be compared to the speed of a stone, which rushes with like fury down the hill, whether it was first put in motion by the arm of a giant or the hand of a boy" (107)—the same analogy that draws our attention to the phenomena of mass, force, and acceleration. But to name his action in this scene would require multiple negatives and redundancies: Ravenswood does not not stop the force (the bull) that would have eliminated the force (Ashton) that is accelerating the historical force that is trying to eliminate him. To not act is also to act, when something is moving fast in someone's direction—as it usually is, in this particular novel.

The calculations, beliefs, even the fatalism of characters merely provide an ironic counterpoint to the momentum of history, of the plot. The novel offers numerous references to legends and prophecies about the downfall of the Ravenswoods, and even invites antiquarians or folklorists to study them. In one moment, Ashton chances to remember a legend about an ancestral Ravenswood and a long-deferred revenge; this thought makes him pause while writing a letter that would have denounced Ravenswood to the authorities, "as if for the purpose of . . . reflecting farther on the consequences of the step which he was about to take, ere yet they became inevitable" (39). Ashton thinks that he is pausing, stopping the action, doing nothing, but he had already set things in motion, and the consequences of those machinations continue to unfold, as he knows and secretly counts on. The creature's limbs will keep moving forward until they are stopped by some external cause, in another strange illustration of both Newton's and Darwin's laws of animal motion. Ashton may think he's acting or refraining from acting in rational ways, but, in the end, he relies on a deeper faith in inertial movement, both to attain his desired end and to absolve himself of responsibility. And he's far from alone in doing so.

The objectively truer experience of history is found, ironically, in proprioception: the internal sensations of the body's muscles and joints, that somewhat unreliable source of balance and perpendicularity, according to Darwin. When Ravenswood joins a hunt that happens to pass by, he loses himself in just these sensations: "The feeling of his own rapid motion, animated the Master of Ravenswood . . . above the recollections of a more painful nature by which he was surrounded." (107). On more than one occasion, Ravenswood tries to displace his conscious awareness by the sensation of his own physical motion. He gives up trying to figure out where he stands, or what the ambient objects are doing, and absorbs himself in this feeling. Appropriately, at the end of the novel, the quicksand seems simply to swallow him up, as another person looks on in disbelief.

Scott's 1824 novel, *Redgauntlet* (subtitled *A Tale of the Eighteenth Century*), most fully builds a plot on the historicity of the body's fibers, and on the sense of proprioception. It uses the inescapability of vection in order to invoke both possibility and impossibility.[34] The novel is about a rebellion that does not happen: Hugh Redgauntlet is attempting again to restore Charles Edward Stuart to the throne, twenty years after the failed Jacobite uprising of 1745. To further his plan, he kidnaps his nephew, Darsie Latimer, who had been brought up by a foster family in Scotland without knowing his true family identity. He forces Darsie to accompany him to the gathering that is supposed to initiate the movement, but ultimately without success. At the end, the old Jacobites whom he has managed to assemble find an excuse to slink away. A representative of the government shows up, but allows Redgauntlet and the aging Pretender he has brought to England to depart peacefully and without punishment, and history seems to have reached an end, with the present government's legitimacy established forever.[35] But the novel's emphasis on vection and vertigo suggests a different understanding.

The actions of both protagonists are presented as movements of the body's fibers that occur independently of volition. Darsie, who longs to know his family identity, starts the novel idly haunting the border between Scotland and England, where he has been forbidden to go. He has no conscious aim as he lingers on the sands of the Solway Firth: "The anxious thoughts which haunt me began to muster in my bosom, and my feet slowly and insensibly approached the river which divided me from the forbidden precincts."[36] Redgauntlet's support of the Stuarts is also explained as a hereditary loyalty that he cannot escape; it is first illustrated by his way of walking. Darsie has been rescued from the quicksand by his uncle without knowing the latter's identity; he hears Redgauntlet pacing in his room all night, and thinks, "His pace . . . seemed to keep time with some current of internal passion, dark, slow, and unchanged. . . . What thoughts may be now corresponding with that furrowed brow, and bearing time with that heavy step!" (44). The family identity of the Redgauntlets consists in a peculiar involuntary activity of the muscles: Darsie first suspects he may be related to Redgauntlet when the latter gives him "one of those extraordinary looks by which he could contort so strangely the wrinkles on his forehead." Darsie instinctively "answered him by a look of the same kind and, catching the reflection of my countenance . . . I started again at the real or imaginary resemblance which my countenance, at that moment, bore" (199–200). Darsie tries to replicate the expression, but finds that it cannot be done voluntarily. Family identity is figured here not as a static resemblance of feature but as a passing, involuntary muscular contraction—a "catenated train of ideas and muscular motions," in Darwin's language. Note, though, that Darsie isn't sure whether it really exists; he has to look in a mirror or judge by someone else's response to his expression. Proprioception, or the feelings produced by one's own muscles and joints, is highly uncertain.

Yet the novel's plot relies to an unusual degree on the feelings of the muscles and joints when standing or moving and the ideas thus elicited—for example, the feeling of standing on stable or unstable ground, the feeling of pushing against something heavy, or of walking up or down an incline. As Darsie lingers on the literally unsettled ground that is the tidal estuary separating Scotland and England, he realizes that the tide is coming in rapidly: "I began a race as fast as I could, feeling, or thinking I felt, each pool of salt water through which I splashed, grow deeper and deeper. At length the surface of the sand did seem considerably more intersected with pool and channels full of water . . . the sands at the same time turned softer" (33). Darsie second-guesses these sensations; indeed, they prove extremely unreliable in conveying any information about position or direction. After he's kidnapped by his uncle, Darsie finds himself bound and being taken across the same estuary to England: "My sensations were horrible . . . I became conscious that I was carried violently forward in some conveyance, with an unequal motion, which gave me much pain. My position was horizontal . . . I opened my eyes, it was to no purpose—all around me was dark. . . . It is with the utmost difficulty that I at length, and gradually, recovered . . . the power of observing external sounds and circumstances" (173). Here again an effort has to be made to turn attention outward, but even this does not provide answers. He feels himself being thrown from side to side and up and down, but doesn't know where he's headed, or how fast he's moving.

This phenomenon (along with the involuntary, supposedly hereditary frown, "the fatal mark of our race," as Redgauntlet calls it [208]), is used to explain the historical, as well as the physiological, identity of this family. The Redgauntlets are cursed by a legendary act of filicide to be on the losing side of every civil conflict in Scotland: "Often making a figure in history, they have been ever in the situation of men striving against both wind and tide, who distinguish themselves by their desperate exertions of strength, and their persevering endurance of toil, but without being able to advance themselves upon their course, by either vigour or resolution" (320). They are fated to be always on the wrong side of history, and yet the only way that they can tell is through this sense of pushing desperately against something. They are not the only ones unable to tell which way history is moving, or whether things are settled. Another character points out to Darsie that the current moment, the peace following the Seven Years' War, is in fact turbulent, full of restlessness and potential conflict. And he argues, though Darsie is doubtful, that traditional feudal allegiances and practices remain powerful forces that can be called upon. We are shown how the latter are entwined with livelihoods and modes of production that are trying to survive. Redgauntlet's followers, who live in the border area, practice traditional methods of salmon-fishing with boat-nets and spears; he incites them to riot against a newer method, "certain improved modes of erecting snares, opening at the advance of the tide, and shutting at the reflux, [by which] many more fish are taken. . . . They complain of these tide-nets . . . as an innovation, and pretend to a right to remove and destroy them by the strong hand" (73). History is the condition in which "both ourselves and the circumjacent objects are in real motion" (Darwin, 234). It is difficult to tell if the world is moving toward or away from you, whether you should push back against it or give in.

The failure of the attempted rebellion is explained in terms of what we might call the undecidability of resistance, of whether there is or is not something stable and solid that we're pushing against. The aging Jacobites that Redgauntlet has assembled find an excuse to abandon their cause. They complain, "'You must not hurry us on too fast, Mr Redgauntlet . . . we will not be driven forward blindfold." Redgauntlet replies, "'Who hurries you, my Lord? Who is it that would drive this meeting forward blindfold?'" (369). But Charles Edward and the Jacobites themselves turn out to be a phantom substance in the eyes of the government. The general who arrives announces that the king will not "offer any obstacle" to their all simply going home peacefully (395). The Pretender walks out to the ship, leaning on Redgauntlet's arm: "For the ground was rough, and he no longer possessed the elasticity of limb and of spirit which [he] had, twenty years before. . . . His adherents followed, looking on the ground" (397). They seem to look at the ground not only out of embarrassment but from the need to project onto it either a quality of stability or, perhaps, of instability, mobility, dynamism. Redgauntlet's final words raise the possibility that "the winning side" (the Hanoverian dynasty) may "in turn become the losing one" (399).

Darwin noted that vertigo on shipboard could be momentarily overcome by a concentrated focus on action and mechanism, the management of sails and ropes. Darsie Latimer, straggling, sinking, uncertain as to who or where he is, finds his attention drawn away from his own situation when, at the beginning of the novel, he watches the fishermen hunting salmon on horseback on the sands of the Solway: "They chased the fish at full gallop, and struck them with their barbed spears, as you see hunters spearing boars in the old tapestry." He forgets himself in admiring the "full command both of . . . horse and weapon" these people display (32). Invoking vertigo as the sensation of historical experience, *Redgauntlet* draws attention prominently to its own action and mechanism as a novel. But instead of counteracting vertigo, its handling and management deepen the sense of disorientation. It begins as an epistolary novel, with letters written by Darsie and his friend to each other. It becomes a journal, and then collapses into a third-person narrative, as if it cannot sustain its form. Is it collapse or development? The novel teems with embedded tales that illustrate the cruelty of the world. It gestures even further back: Darsie, while under his uncle's control, is dressed in a lady's riding habit and mask, a captive damsel who will be rescued by a quixote.

But the future of the novel can also be glimpsed beneath the surface. In a subplot that is given equal importance until it is swallowed up by Darsie's, his friend, Alan Fairford, embarks on his career as a lawyer. For his first case, Alan is given a legendarily complex, massive, and tangled lawsuit, whose half-insane plaintiff, Peter Peebles, also figures in the plot. But just at the point when Alan is on the verge of successfully untangling it, he receives news of Darsie's kidnapping, runs out of court, sets off to find his friend, and leaves the suit unsettled. Perhaps it has to wait for Dickens and *Bleak House* to finally dissolve. Alan's experience pursuing and trying to rescue his friend, though, is also largely one of intense vertigo. He goes by ship, where he suffers terrible "sickness and nausea" (273), then on horseback, where he cannot even sit upright: he is "tormented by racking pain . . . which made the rough trot of the horse torture to him . . . his own bodily distress was now so great and engrossing, that to think of his situation was impossible, even if he could have mended it by doing so" (287). He is wholly engrossed in his own subjective sensations. Carrying a sealed letter to an undisclosed destination, Alan may be a captive, or a rescuer. He is not sure. Movement is constant, but its speed, direction, and meaning are a matter of interpretation: "Their course was inland; but in what direction, Alan had no means of ascertaining" (287). In a larger sense as well, the direction of history cannot be ascertained. Alan is lectured by his ship's world-weary pilot, who believes him to be part of Redgauntlet's plot: "All these rackets and riots that you think are trending your way, have no relation at all to your interest . . . because the London people are roaring about for some pinches of their own, [Redgauntlet] thinks to win them to his turn with a wet finger. And he gets encouragement from some, because they want a spell of money from him; and from others, because they fought for the cause once . . . and others, because they have nothing to lose; and others, because they are discontented fools" (281). What this man describes in admirable sociological detail is historical vection, the condition of not being able to distinguish history's movements from your own. He is also laying out once again the open, unsettled condition of history as the ground of plot. All that Alan can really feel as he makes his way toward

Darsie is the uneven topography of the land, and the sensation of being in a "labyrinth" (287), but that is more than enough for a plot to be built on.

The objective obstacles that novelistic plots focus on can be known, however precariously, by the sensation of pushing against them or sinking into them—not by the outcome alone. The historical condition of being "unsettled" may well be projected, created by the uncertainty of muscular input. But the stasis and solidity of the external world are also projected: the fictional, physiological mechanism that enables us to move through the world without falling constantly. For Darwin the external world was the world of material objects; for Scott and for the realist novel after him, that world is now also conceived as history, as a shifting, elusive configuration, sometimes hopeful and sometimes vengeful. "What was fristed [postponed] wasna forgiven," as another character in this novel states (117), a blind fiddler who feels his way across the country. We must wait to see what happens next. Plot will need to cultivate its own "capillary movements" in attending to those of history.[37]

In an important twentieth-century critical treatment of realism, Erich Auerbach argued that realism, in addition to giving up the classical separation of styles, had to adopt a certain approach to history. History itself had to come to possess a certain inner space or hidden necessity. This idea, he argues, emerged in late eighteenth-century Germany as "a sense of historical dynamics, of the incomparability of historical phenomena and of their constant inner mobility."[38] This sense, which, like Friedrich Meinecke he calls historism, "laid the foundation of modern realism."[39] Historicism views not only the past but "the present . . . as animated by inner forces and in a constant state of development."[40] But the German writers of this period, and Goethe in particular, he notes with disappointment, rejected such a way of representing the present. "The dynamics of opposing social forces and the economic substratum . . . civic unrest . . . these are things which [Goethe] dislikes, and therefore he turns his back on them."[41] But filling the gap between Goethe and Marx, as this chapter has tried to show, Scott's historical novels robustly embody the historicism that Auerbach describes. They explore the dynamic forces that can be sensed, correctly or incorrectly, in any historical moment, but they do so in a way that the novel is now best equipped to do. As natural philosophers like Erasmus Darwin show, humans are organically ill-equipped for adequate comprehension—or even accurate sensation of the movements happening within and around them. Though history never stops moving, they cannot tell which way or how fast it's moving. Curiously, this phenomenon of vection comes to underwrite a belief in history's objectivity.

The fact that historical change eludes the sensory, cognitive, emotional, and moral grasp of characters can be taken as proof that it occurs independently of their agency. History emerges, then, through a sort of negative representation in addition to its actual depiction. It makes itself felt as a complex, objective set of movements and forces that characters do not intend, cannot grasp, apprehend, or steer. And a deeper model of plot emerges from the failure of different kinds of movements to correspond or to keep pace with each other. Motion within the mind, or as experienced by the senses, is especially prone to outpace, to lag behind, or generally fail to correspond to the objective movements that seem to be occurring in the external world. History comes to be conceived as a complex set of forces registered through the body: it pushes and pulls, smooths the way or trips you up. If you look closely enough, or even if you just try to move through your life, everything is in motion, or appears to be.

EPILOGUE

Plot, History, and Totality in Zadie Smith's *White Teeth*

The movements vividly evoked in Scott's novels included those that resulted in the expansion of Britain's empire and then its dissolution and transformation in the second half of the twentieth century.[1] In this brief epilogue, I'd like to look at a novel published after all this has happened, at a time when realism, too, had been consigned to the past. My purpose is to explore the relevance of my argument beyond the historical coordinates of the long eighteenth century. I have suggested that the plots of novels and natural philosophy in this period both explored the conditions in which certain kinds of events and event-sequences signify. They found ways to use carefully contrived experiences to investigate the composition of the world. Moreover, I situated the novel within the history of objectivity as a concept whose definition shifted between the eighteenth and nineteenth centuries. How do these connections between the novel and natural philosophy change? Do they disappear? To give a full answer to these questions is beyond the scope of this study. Fortunately, other scholars have already produced important studies of the manifold connections between science and the novel in other periods.[2] Taking what I hope is not unfair advantage of the epilogue's detached form, I will examine Zadie Smith's *White Teeth* (2000) to see how plot has fared in a later moment. While I cannot rest an argument about the twentieth- or twenty-first-century novel on one example, this work was held up by critics as perhaps embodying a certain stage in the genre's history: "the contemporary idea of the 'big, ambitious novel.'"[3] This distinction it seems to owe to its elaborate and rather obtrusive plotting. James Wood wrote, "The contemporary novel is a perpetual-motion machine that appears to have been embarrassed into velocity. . . . Stories and sub-stories sprout on every page."[4] Does this plotting serve a purpose? Is it the novel that is embarrassed by its excessive plotting or the reader? And what does that response suggest about the kind of objectivity that a novel is still expected to possess? How is that busy plotting related to the novel's realism?[5]

In a way familiar to readers of eighteenth- and nineteenth-century novels, Smith's plot makes uninhibited use of accidents, coincidences, small objects, and stories within stories. Minor characters reemerge to suddenly take on a significant role. There are long-kept secrets. Trajectories converge. This is far from the Defoe-type of plot that Scott admired for its looseness, its heap-of-paving-stones quality. Smith's novel stands out, however, not only for the virtuosity of its plot but for what it aims to do through such means. "It is not the writer's job," Smith has said elsewhere, "to tell us how somebody felt about something, it's to tell us how the world works."[6] The latter can best be achieved, it seems, through the foldings and unfoldings of plot. Plot in this novel acquires a certain awareness of itself, I want to suggest, in relation to the history of empire and the science of genetics. It takes up without embarrassment the epic function of plot, which is to present a detailed external world, extended both in space and time, that largely defeats or frustrates the aims of the characters within it.

The many characters in this novel, located in different classes, religions, and professions, all try their best to plan and to predict. They spend much of their time imagining both short-term and long-term consequences, future chains of events. Jehovah's Witnesses industriously plan for the end of the world, undeterred by past nonhappenings. Scientists create a transgenic mouse whose life, illnesses, and death are all precisely planned almost to the day. Immigrant parents try to determine their children's future characters. Nearly all of these plans will be defeated, but it can't be said that chance or accident is solely responsible. Rather, the plot draws attention to itself as the agent. Yet to do so, it also leans noticeably on the alibis of the material world, or on the properties of that world that were so clearly described by eighteenth-century natural philosophers. In the words of Adam Smith, for example, "The real inertness of matter" lies "not in an aversion to motion, or in a propensity to rest, but in a power of continuing indifferently either at rest or in motion, and of resisting, with a certain force, whatever endeavoured to change its state from the one to the other."[7] This account could almost serve to summarize many parts of the plot in Smith's novel. The following forms of praxis seem to be available to its characters: to plan, to predict, to protest, and to resist. Most of the actions, whether performed by children, grandmothers, or scientists, fall into these categories. But in all these cases, we have the sense of watching not actions taken by individual characters so much as trajectories that have already begun. They are often collective trajectories. The initial precipitating causes are tangled, murky, complex. The point is not to disentangle those causes but to see where and how the thing falls. The suspense resides there, between the moment of tossing the coin in the air and seeing how it lands: an event that occurs, like so many others in this novel, twice. It is clearly an illustration of probability, indeed, the classic trope of probability. But what matters the most in both instances, when Archie Jones tosses his coin, is not whether it lands tails up or heads, but rather the trajectory of the coin's motion, and the exact location in the novel's physical setting where it comes to rest.

The plot of *White Teeth* is built around three men, with a fourth man deep in the background. Barely named when he first appears and barely mentioned after that, that fourth man, the eugenicist Marc-Pierre Perret, is revealed at the end to have been the causal node to which nearly everything else is linked. The Englishman Archie Jones and the Bengali Samad Iqbal meet as very young men in the British army; they have arrived in Europe just after the conclusion of the Second World War; their only remaining task is to take into custody a scientist who has been working for the Germans. Iqbal proposes to "test" Jones, to see if he can kill this man: "Call tonight. . . . An experiment . . . I want to know what kind of man you are, Jones. I want to know what you are capable of."[8] The stake is higher than Archie's individual character, however. Iqbal aims to understand through this test how the British empire came to be a fact, "How your lot ever conquered my lot" (101). Jones leads Perret away, and a gunshot is heard, but the scene is not revealed until the final pages of the novel. We then see that Jones flips a coin after Perret argues and pleads for his life, but the coin lands behind him, forcing him to turn, allowing Perret to grab the gun and shoot Jones in the leg instead. Perret survives, we learn at the end, to become the mentor of the third man, Marcus Chalfen, a biologist who plays a decisive role in the development of the Joneses' and Iqbals' children. Chalfen also develops and patents the "FutureMouse," a mouse that has been genetically programmed to develop tumors at specific times and dates, and to die at the end of 1999.

Smith's novel is replete with dates, historical events, and science. The "FutureMouse" is based on the actual "Oncomouse," a type of mouse whose embryo was engineered to carry a cancer-related gene; it was patented in 1988 and trademarked in 1990.[9] Since it targets the most intimate material composition of living things, genetic engineering is an apt partner for the realist novel.

Smith's plot, however, hardly defers to the authority of either science or historiography. Rather, it embeds both within a larger, teeming landscape of commensurate projects that aim to find meaning or assert control over what all subjects keenly perceive as a wayward materiality. The plot establishes humorous but also serious equivalences between Chalfen's science and the fundamentalism of second-generation British Muslims. Beneath the surface of its satire, it suggests that science has benefited from historical atrocity, at worst, and, at best, that it legitimates social prejudice.[10] History, likewise, is dissected into multiple questionable modes. The purely "external" form is illustrated by the bus ticket we see reproduced at the end of the novel: "On: 31/12/1992/London Transport Buses/Route 98/ From: Willesden Lane/To: Trafalgar Square/ At: 17:35" (422)—a meaningless time stamp. Even the fall of the Berlin Wall, the "end of history," is something that merely happens: "November 10, 1989. A wall is coming down. It was something to do with history. It was *an historic occasion*. No one really knew quite who had put it up or who was tearing it down or whether this was good, bad, or something else; no one knew how tall it was, how long it was. . . . It was a Thursday night . . . everybody was watching history on TV" (197). But history is more often the subject of inconclusive debate. Iqbal never tires of asserting, over the objections of others, that his great-grandfather, Mangal Pandey, had begun the Indian mutiny of 1857. History, or more particularly, historical causalities become mere aspects of subjective identity, something to appropriate, cling to, or repossess.

The novel states (twice): "Every moment happens twice: inside and outside, and they are two different histories" (299, 441). This observation seems to refer to the uneven distribution of power: some have the power to act, while others have none. There may well be a divergence between history as I experienced it, or as I wish it had happened, and the history that actually occurred. But this novel's plot engineers their collision, subsuming the difference between them. It also undertakes to make events happen twice, as histories that are different and yet the same. An example is found in Glenard Oaks, the school that the Iqbal, Jones, and Chalfen children attend. The original building was built as an "experiment" (254) by Sir Edmund Glenard, a Jamaican tobacco plantation owner who sent "three hundred Jamaicans to North London" to work there "packaging Sir Edmund's cigarettes and taking general instruction from the Englishmen in the evening" (254). It is still a place dominated by tobacco. The school is populated by multiracial gangs of "twelve-, thirteen-year-old chain-smokers. . . . Smoking was their answer to the universe . . . their raison d'être . . . everybody, everybody smoking . . .

chuffing on them like little chimneys" (242–43). Sir Edmund is also implicated in the Jones family history: the plot has him "crushed to death by a toppled marble madonna" during the Kingston earthquake of 1907, in the act of sexually assaulting the Jamaican great-grandmother of Irie, Archie's daughter. In 1990, another earthquake occurs when Irie and Millat Iqbal are caught smoking at school, which is still technically illegal, during a raid: "She knew the day and time of the quake (today, two-thirty), she knew its power . . . and she knew who was likely to fall victim" (244). As a substitute for detention, the two teenagers are made to go after school to the Chalfens' house to receive mentoring and tutoring in math and biology. The headmaster imagines it as a model: "Bringing children of disadvantaged or minority backgrounds into contact with kids who might have something to offer them" (256). But it had been done before. And the consequences of this second iteration of colonial history will shape the rest of the plot, pushing Irie into the Chalfens' world, Millat into a Muslim group ("KEVIN"), and the Chalfens' son, Joshua, into a radical animal rights organization ("FATE"). The two latter groups converge in purpose and in person to undo Chalfen and emancipate the "Future-Mouse" at its triumphant public unveiling. On the same occasion, Jones will save Perret's life for a second time, when the somewhat befuddled Millat shoots at the scientist: "So Archie is there, there in the trajectory of the bullet, about to do something unusual, even for TV: save the same man twice and with no more reason or rhyme than the first time" (447).

Archie Jones's body comes between the bullet and the man it's heading toward: "He is there between Millat Iqbal's decision and his target . . . like the split-second intervention of memory or regret" (442). The interposition of his leg performs the same function in the plot as a change of heart or intention. As he falls, the trajectory of his body produces another significant effect: "Everybody in the room watches in horror as he takes it in the thigh . . . and falls right through the mouse's glass box" (447). In both cases, Jones's body behaves as one of the many intermediate causes manipulated by the plot. *White Teeth* avails itself fearlessly of what Defoe had called "the Treasury of Immediate Cause," all those material things whose proximity or conjoined movements result in change.[11] As if to remind us, the plot even features the great storm of October 1987, "the worst storm since the Great Storm of 1703," and a notable failure in weather forecasting.[12] Luckier than Defoe's scrivener, Archie moves just in time to avoid "a mammoth tree . . . [that] took one tottering step forward, swooned, and collapsed; through the guttering, through the storm windows, knocked over a gas lamp, and then landed in an absence that was Archie-shaped, for he had just left it" (188). At the start of the novel, Archie tries to commit suicide, parked on a street; his act is thwarted by a chain of causes that begins with a flock of pigeons landing on a roof and ends with the halal butcher in front of whose shop he had parked demanding that he move his car. The novel coyly asserts that the ultimate cause cannot be known: "The position of the planets, the music of the spheres, the flap of a tiger moth's diaphanous wings. . . . Somewhere, somehow, by somebody, it had been decided that he would live" (4). But intermediate causes, as natural phenomena, obey universal laws: "To the same natural effects we must, as far as possible, assign the same causes. As to respiration in a man and in a beast; the descent of stones in Europe and in America; the light of our culinary fire and of the sun; the reflection of light in the earth and in the planets."[13]

To see the varied elements of plot in this way, though, requires abstracting from particulars. Historical differences between events have to be erased, as they become things for the plot to think with. In one scene, Millat Iqbal and his twin, Magid, engage in just such an exercise. Magid, who had been sent to Bangladesh by his father as a child to become a devout Muslim, returns as a rationalist devoted to Western thought, science, and English law. The brothers meet in order to argue, using every prop in the room as models, diagrams, and substitutes in order to consider "every debated principle, every contested belief" (383). But this collision fails to change the course of either twin: "They left . . . as they had entered . . . unable to waver from their course or in any way to change their separate, dangerous trajectories" (384). Those trajectories have already been set, it seems, by the history of empire as well as the history of science and of the novel. "Do we *really* have a choice?" Irie asks, when given the option to be tutored by the Chalfens. She answers her own question, "We don't have a choice" (255). As another character reflects, everyone and everything is already "involved": "Involved . . . is just a consequence of living, a consequence of occupation and immigration, of empires and expansion . . . one becomes involved . . . they are not *wanting* this, they are not *willing* it—they are just involved, see? . . . An enormous web" (363, original emphasis).

But it is not enough just to ascribe this involvement to something called history. In this novel, involvement follows the paths or takes the shapes of the structures and infrastructures offered to our notice. Those range from traffic circles to the layouts of parks and public spaces, from the folds in the paper leaflets that Archie devises, working for a direct mail advertiser, to the kind of division that Jones and Iqbal had been assigned to in the war: "One of the specialist divisions . . . laying bridges, creating passages for battle, creating routes. . . . Their job was not so much to fight the war as to make sure it ran smoothly" (74). The novel's finale begins by tracing at length several simultaneous journeys made into central London via van, tube, and bus. Not only is the bus ticket reproduced, but this notice at the subway station as well: "Signaling problems at Baker Street/No Southbound Jubilee Line Trains from Baker Street/Customers are advised to change onto the Metropolitan Line at Finchley Road/Or change at Baker Street on to the Bakerloo" (412). The infrastructure, along with the novel's governing conceit of "root canals," invokes the nature of plot itself. These are the actions that the plot sees itself performing: to connect, to gather, to enable trajectories and collisions. Its job is to fold or to connect these events so that their existence as a mass of unrelated particulars is overcome. Plot in this novel seems to negate chance as firmly as it does subjective will. What matters is not the probability of an event but the actual consequence, and the consequence is framed by the deep channels and pathways of plot.

In the novel, Archie tosses his coin twice, on two occasions separated in time and in the narrative discourse. Both times, the identical description is given of the event: "The coin rose and flipped as a coin would rise and flip every time in a perfect world flashing its light and then revealing its dark enough times to mesmerize a man. Then, at some point in its triumphant ascension, it began to arc, and the arc went wrong, and Archibald realized that it was not coming back to him at all but going behind him, a fair way behind him, and he turned . . ." (377, 447). The chronologically first time, this event results in the escape of Perret. The chronologically second time, which comes earlier in the novel's discourse, there is a different outcome: "He turned with the others to watch it complete an elegant swoop toward the pinball machine and somersault straight into the slot. Immediately the huge old beast lit up; the ball shot off and began its chaotic, noisy course around a labyrinth of swinging doors, automatic bats, tubes, and ringing bells until, with no one to assist it, no one to direct it, it gave up the ghost and dropped back into the swallowing hole" (377).

The ball's movement occurs "with no one to assist it, no one to direct it." It illustrates something like pure plot objectivity. Its movement sets off a chain of gratifying events that occur not in a "perfect world," but in a world that is regular but imperfect and therefore, by implication, real. Plot: "the huge old beast," the totalizing machine.

NOTES

Chapter 1

1. William Godwin, "Of History and Romance," in *Caleb Williams*, ed. Maurice Hindle (London: Penguin, 1988), 369–73, 372. Hindle notes that the essay was unpublished but probably written in 1797.

2. Ibid.

3. The phrases echo Newton's own description of his work, as the second chapter of this book addresses in more detail. Godwin does not specify here whether "that principle" of gravity depends on God or works on its own. On the concept of the system, see Clifford Siskin, "Mediated Enlightenment: The System of the World," in *This Is Enlightenment*, ed. Clifford Siskin and William Warner (Chicago: University of Chicago Press, 2010), 164–72; and Clifford Siskin, *System: The Shaping of Modern Knowledge* (Cambridge, MA: MIT Press, 2016).

4. See Penny Fielding, "'No Such Thing as Action': William Godwin, the Decision, and the Secret," *Novel: A Forum on Fiction* 42, no. 3 (2009): 380–86.

5. Usually, this novel (full original title: *Things as They Are; or, The Adventures of Caleb Williams* [1794]) is referred to by critics as *Caleb Williams*, an abbreviated version of its subtitle; in itself, this reveals the tendency of critics to identify the novel with an individual subject.

6. Arguably, even a wish, hope, or fear would count as an objective mental event for the purposes of the fiction. For a helpful discussion of twentieth-century philosophies of the event, as well as their relation to the modernist novel, see Michael Sayeau, *Against the Event* (Oxford: Oxford University Press, 2013), 1–48. On eventfulness in relation to "tellability" or "narratability," see Peter Huhn, *Eventfulness in British Fiction* (New York: De Gruyter, 2010). Huhn defines eventfulness as "departure from a schematic pattern or script activated in the text" (6) and notes that it is a "scalar category" (3).

7. Powerful feelings are usually elicited by such discoveries. On recognition scenes, see Terence Cave, *Recognitions* (New York: Oxford University Press, 1988). Interestingly, one such feeling can be disbelief: Scott's first hero, Waverley, feels exactly this emotion when he finds himself caught up in the 1745 Jacobite rebellion. Walter Scott, *Waverley*, ed. Claire Lamont (Oxford: Oxford University Press, 1986).

8. Objectivity in this sense as a dimension of the novel's plot overlaps with probability, verisimilitude, questions of truth, all questions that have been well investigated by critics from Aristotle, to neoclassical critics of the seventeenth and eighteenth centuries, to Bacon, and theorists of probability; see, for example, Michael McKeon, *Origins of the English Novel* (Baltimore: Johns Hopkins University Press, 1987); Jesse Molesworth, *Chance and the Eighteenth-Century Novel* (Cambridge: Cambridge University Press, 2010). Marie-Laure Ryan has usefully named the "textual actual world" that which "exists absolutely in the semantic universe of the text, as

opposed to what exists in the minds of characters. See her *Possible Worlds, Artificial Intelligence, and Narrative Theory* (Bloomington: Indiana University Press, 1991), 112.

9. David Herman, *Story Logic* (Lincoln: University of Nebraska Press, 2002), 2. On alternative possible worlds, see Ryan, *Possible Worlds*. See also Karin Kukkonen, *4E Cognition and Eighteenth-Century Fiction* (New York: Oxford University Press, 2019); and Karin Kukkonen, *Probability Designs: Literature and Predictive Processing* (New York: Oxford University Press, 2020), for a fascinating account of how readers may reconstruct the cues given by narratives into firsthand experience, as well as how important a role prediction plays, when it comes to experiencing plot. My approach will be somewhat different, as this chapter tries to show, in aiming at a greater historical specificity.

10. Herman, *Story Logic*, 16. On alternative possible worlds, see Ryan, *Possible Worlds*; and Lubomír Doležel, *Possible Worlds of Fiction and History* (Baltimore: Johns Hopkins University Press, 2010).

11. In the earlier part of the century, this often meant claiming to be literally true or factual, as was the case with Defoe. See McKeon, *Origins of the English Novel*, as well as J. Paul Hunter, *Before Novels* (New York: Norton, 1990). I will not discuss other kinds of fictional narratives that made other sorts of claims, such as Oriental tales, "it-narratives," or Gothic novellas. For a comprehensive overview, see the essays in *English and British Fiction, 1750–1820*, ed. Peter Garside and Karen O'Brien (Oxford: Oxford University Press, 2015).

12. In this sense, all events in realist plots, I would argue, can be understood through the model of metalepsis described in Elaine Freedgood's recent study, *Worlds Enough* (Princeton: Princeton University Press, 2019).

13. Most influentially in Ian Watt's *The Rise of the Novel* (Berkeley: University of California Press, 1957), discussed below.

14. Peter Dear, *Discipline and Experience* (Chicago: University of Chicago Press, 1995), 4.

15. *Aristotle's "Poetics,"* trans. James Hutton (New York: Norton, 1982), 54.

16. This may well apply to narrative plots in general; think of how often the same event occurs three times in folktales, for example.

17. Joachim Jung, cited in Dear, *Discipline and Experience*, 221; Isaac Barrow, cited in Dear, *Discipline and Experience*, 222.

18. As Nicholas Paige argues of "the complexity, the counterintuitiveness, of the realist operation, and its difference with regard to earlier novels' assertions of truth . . . one cannot literally copy abstractions like 'men' and 'society'" (*Before Fiction* [Philadelphia: University of Pennsylvania Press, 2011], 17).

19. On the importance of these developments, see Lorraine Daston, *Classical Probability in the Enlightenment* (Princeton: Princeton University Press, 1988); and Douglas Patey, *Probability and Literary Form* (Cambridge: Cambridge University Press, 1984).

20. Dear, *Discipline and Experience*, 3–4.

21. Ibid., 6.

22. Jerome Bruner, "The Narrative Construction of Reality," *Critical Inquiry* 18 (1991): 1–21, 11.

23. Ibid., 13–14. Bruner gives two examples: when Joyce mentions St. Stephen's Green in Ulysses, it becomes "an invented referent not entirely of the meanings imparted by the real place, just as a story that requires a 'betrayal' as one of its constituent functions can convert an ordinarily mundane event into something that seems compellingly like a betrayal" (13–14).

24. Ibid.

25. See the essays in *The Oxford Handbook of Causation*, ed. Helen Beebee, Christopher Hitchcock, and Peter Menzies (Oxford: Oxford University Press, 2009).

26. Latour's method could be described as collapsing the distinction between describing setting and plot; plot consists in describing the various stages of movement and embodiment of, for example, soil samples in the Amazon forest; every step is transformative, with no events or agents privileged over others. See Bruno Latour, *Pandora's Hope: Essays on the Reality of Science Studies* (Cambridge, MA: Harvard University Press, 1999).

27. As Bruno Latour has argued in *Reassembling the Social* (Oxford: Oxford University Press, 2005), "the social" dissolves into a network of circulating forces and associated agents, themselves often impelled more by a host of small circumstances than by a single, concerted act of the will. Recall that in his study of Balzac's *Sarrasine*, Roland Barthes's list of plot or *proairetic* sequences includes several that are named "door." The plot sequence that he names "amorous outing" is analyzed into "1: to get into the same carriage . . . 6: to assist in alighting from the carriage" (*S/Z*, trans. Richard Miller [New York: Hill and Wang, 1974], 255–59).

28. On related ways to define eventfulness, see Huhn's discussion of Jurij Lotman's idea of an underlying "semantic field" internally divided by borders (*Eventfulness in British Fiction*, 7).

29. Jane Austen, *Pride and Prejudice*, ed. James Kinsley (Oxford: Oxford University Press, 2004), 280–98.

30. Arguably, the role played by God in natural theology transcended that of mundane causation; see the discussion in Chapter 2 below.

31. Brian Richardson argues convincingly that "the system of causation in a narrative is as basic a component of its setting as are the related elements of time and space"; see *Unlikely Stories* (Newark: University of Delaware Press, 1997), 61. The novels that I consider fall within what he describes as a "naturalistic universe" (67). Richardson points out that novels with indeterminate settings in time and space also lack apparent causal laws (79).

32. Lorraine Daston and Peter Galison, *Objectivity* (New York: Zone Books, 2007), 70.

33. Ibid., 197, 184–85. See George Levine, *Dying to Know* (Chicago: University of Chicago Press, 2002), for a more elaborate account that relates such an ideal to Victorian literature.

34. Daston and Galison, *Objectivity*, 17.

35. See Nancy Armstrong, *How Novels Think* (New York: Columbia University Press, 2005).

36. Daston and Galison, *Objectivity*, 201–3.

37. Michel Foucault, *The Order of Things* (New York: Vintage, 1983), 219.

38. Ibid., 239–40.

39. Foucault argues that Kant's critical philosophy "brings out the metaphysical dimension" that had been concealed within the classical ideology of representation notably formulated by Destutt de Tracy (*The Order of Things*, 241–43).

40. There are surely many factors involved besides Kant's critical philosophy; we might note the expanded activity and industrial application of natural philosophy in the final third of the eighteenth century, not to mention global revolutions and struggles for power.

41. See Christine Lehleiter, ed., *Fact and Fiction: Literary and Scientific Cultures in Germany and Britain* (Toronto: University of Toronto Press, 2016); John Heilbron, "Natural Philosophy," in *Wrestling with Nature: From Omens to Science*, ed. Peter Harrison, Ronald L. Numbers, and Michael H. Shank (Chicago: University of Chicago Press, 2011), 173–201. In the chapters below, I cite other important works that examine the development of science in the Romantic

era. On the novel, see in particular Homer Obed Brown, *Institutions of the English Novel from Defoe to Scott* (Philadelphia: University of Pennsylvania Press, 1997).

42. Clara Reeve, *The Progress of Romance*, 2 vols. (Colchester: W. Keymer, 1785), 2:7.

43. Ibid., 2:38–39.

44. Ibid., 1:111.

45. Daston and Galison, *Objectivity*, 207.

46. See Claudia Johnson, "'Let Me Make the Novels of a Country': Barbauld's *The British Novelists*," *Novel: A Forum on Fiction* 34, no. 2 (2001): 163–79. Johnson offers an important comparison of Barbauld's collection with Scott's, along the lines of gender and politics, as well as nonrealist modes of fiction.

47. Brown, *Institutions of the English Novel*, 183.

48. Anna Laetitia Barbauld, "On the Origin and Progress of Novel-Writing," *The British Novelists*, vol. 1 (London: Rivington, 1820), 15.

49. Ibid.

50. Walter Scott, "Daniel Defoe," in *Sir Walter Scott on Novelists and Fiction*, ed. Ioan Williams (New York: Barnes and Noble, 1968), 164–83, 172. A letter to Ballantyne suggests that Defoe might have been included as the eleventh volume of the Ballantyne series (164).

51. Ibid., 172.

52. Ibid.

53. Ibid., 180.

54. Ibid.

55. Brown, *Institutions of the English Novel*, 187–88.

56. Walter Scott, "Emma," in Williams, *Sir Walter Scott on Novelists and Fiction*, 225–36, 235.

57. Ibid.

58. Watt, *Rise of the Novel*, 13.

59. In the capitalist order, Watt writes, "The effective entity . . . was . . . the individual: he [*sic*] alone was primarily responsible for determining his own economic, social, political, and religious roles" (61).

60. But note that Watt himself posits an objective "principle": "The importance of the plot is in inverse proportion to that of character" (279). He still identifies objectivity with a certain way of seeing or describing, however, as in the following: "Fanny Burney and Jane Austen followed Fielding in adopting a more detached attitude to their narrative material, and in evaluating it from a comic and objective point of view" (296). Watt stands in the modernist tradition, which can be linked to the discussions of twentieth-century objectivity in Daston and Galison, *Objectivity*.

61. Peter Brooks, *Reading for the Plot* (New York: Vintage, 1985), 37–39. Originally published by Alfred A. Knopf, Inc., in 1984.

62. Nancy Armstrong, *Desire and Domestic Fiction* (New York: Oxford University Press, 1987), 23.

63. Daston and Galison, *Objectivity*, 259.

64. See Lee T. Lemon and Marion J. Reis, eds., *Russian Formalist Criticism* (Lincoln: University of Nebraska Press, 2012); on questions of scale in narrative theory, see Yoon Sun Lee, "Questions of Scale: Narrative Theory and Literary History," in *Cambridge Companion to Narrative Theory*, ed. Matthew Garrett (Cambridge: Cambridge University Press, 2018), 29–45.

65. Fredric Jameson, for example, uses the semiotic rectangle to account for what happens in a novel: see the chapter on Conrad, "Romance and Reification," in Fredric Jameson,

The Political Unconscious (Ithaca: Cornell University Press, 1981), 206–80. Jameson revealingly describes the plot of Conrad's *Lord Jim* as a "slow analytical rotation" (271).

66. Vladimir Propp, *Morphology of the Folktale*, trans. Laurence Scott, ed. Louis A. Wagner (Austin: University of Texas Press, 2015).

67. Ibid., 62–63. Readers familiar with Fielding's novels will no doubt notice how well this applies to his endings.

68. Ibid., 63.

69. See ibid., 99. Propp notes "that decomposition into components is, in general, extremely important for any science" (99).

70. Ibid., 22.

71. See Tzvetan Todorov, "The Grammar of Narrative," in *The Poetics of Prose*, trans. Richard Howard (Ithaca: Cornell University Press, 1977), 108–19; Gerald Prince, *A Grammar of Stories* (The Hague: Mouton, 1973). Some recent work continues this project; Robert Belknap, for example, notes that "even this smallest incident [in a plot] has an internal structure of elements that, taken separately, are not incidents any more than hydrogen and oxygen separately are water"—a structure summarized as "a situation, a need, and an action." See Robert Belknap, *Plots* (New York: Columbia University Press, 2016), 34. A similar model is proposed by Emma Kafalenos in *Narrative Causalities* (Columbus: Ohio State University Press, 2006).

72. Roland Barthes, "An Introduction to the Structural Analysis of Narrative," in *A Barthes Reader*, ed. Susan Sontag (New York: Hill and Wang, 1982), 251–95, 266.

73. Roland Barthes, "The Reality Effect," in *The Rustle of Language*, trans. Richard Howard (Berkeley: University of California Press, 1986), 141–48, 148, 146. For an important alternative account of what the realist novel does, see Freedgood, *Worlds Enough*.

74. Some have considered this question by looking at digressions, episodes, or other features defined in formal terms, for example; others have examined descriptions as ways in which a novel expands itself. See Matthew Garrett, *Episodic Poetics* (New York: Oxford University Press, 2014); Cynthia Wall, *The Prose of Things* (Chicago: University of Chicago Press, 2006).

75. In treating novels, critics tend to give more attention to the social dimension, even using them to classify types of plot: the marriage plot, or the plot of education, for example. On the marriage plot, see Armstrong, *Desire and Domestic Fiction*; Kathy Psomiades, "The Marriage Plot in Theory," *Novel: A Forum on Fiction* 43, no. 1 (2010): 53–59; Jill Gavan and Elsie Michie, eds., *Replotting Marriage in Nineteenth-Century British Literature* (Columbus: Ohio State University Press, 2018). On the plot of education and of development, see, for example, Franco Moretti, *The Way of the World* (London: Verso, 2000); Jed Esty, *Unseasonable Youth: Modernism, Colonialism, and the Fiction of Development* (New York: Oxford University Press, 2012); Ian Duncan, *Human Forms* (Princeton: Princeton University Press, 2019); on the food plot, see Michael Parrish Lee, *The Food Plot in the Nineteenth-Century British Novel* (London: Palgrave Macmillan, 2016).

76. G. W. F. Hegel, *Aesthetics*, trans. T. M. Knox, 2 vols. (Oxford: Clarendon, 1975), 2:1081. Hegel's lectures were delivered from 1818 until 1829; they were published later from students' notes and a manuscript that has not been found.

77. Ibid., 2:1038.

78. Ibid., 2:1038–39.

79. Ibid., 2:1037. The situation or field, something like what Kenneth Burke calls the scene, or material conditions of action, takes center stage in epic. See *A Grammar of Motives* (New York: Prentice-Hall, 1945).

80. Hegel, *Aesthetics*, 2:1051.

81. Ibid., 2:1160. I will not emphasize in the text the use of the male pronoun, but wish to point out how different the quotation appears and what different associations are summoned if the gender is changed. On the intricate and far-reaching relations between gender, sexuality, plot, and narration, see Susan Lanser, *The Sexuality of History* (Chicago: University of Chicago Press, 2014).

82. Hegel, *Aesthetics*, 2:1063–64. For a valuable perspective on the relation between event and action, see Jonathan Kramnick, *Actions and Objects from Hobbes to Richardson* (Stanford: Stanford University Press, 2010),

83. Hegel, *Aesthetics*, 2:1064.

84. Barbauld, "Origin and Progress of Novel-Writing," 1:1.

85. On this correspondence, see Richard Littlejohns, "The Discussion Between Goethe and Schiller on the Epic and Dramatic and Its Relevance to *Faust*," *Neophilologus* 71, no. 3 (1987): 388–401.

86. Hegel, *Aesthetics*, 2:1069.

87. Ibid., 2:1086.

88. Johann Wolfgang von Goethe, *Wilhelm Meister's Apprenticeship*, ed. and trans. Eric A. Blackall (Princeton: Princeton University Press, 1995), 185–86.

89. Hegel, *Aesthetics*, 2:1046.

90. Ibid., 2:1053.

91. Georg Lukács, *The Theory of the Novel*, trans. Anna Bostock (Cambridge, MA: MIT Press, 1983), 89.

92. Ibid., 98.

93. Ibid., 125.

94. Ibid., 124–28. This is how he describes the "true epic objectivity" of the "novel of disillusionment."

95. Georg Lukács, *The Historical Novel*, trans. Hannah Mitchell and Stanley Mitchell (Lincoln: University of Nebraska Press, 1983), 123–24.

96. Ibid., 123–24, original emphasis.

97. Ibid., 149–50.

98. Ibid., 146.

99. Hegel, *Aesthetics*, 2:1060.

100. See Sharon Marcus, "Erich Auerbach's *Mimesis* and the Value of Scale," *MLQ* 77, no. 3 (2016): 297–319. Marcus draws attention to Auerbach's "power to shift scales," particularly through toggling between description and interpretation (311–13).

101. See Marshall Brown, "Plan vs. Plot," *Stanford Literature Review* 4, no. 1 (1987): 103–36.

102. Erich Auerbach, *Mimesis* (Princeton: Princeton University Press, 1974), 548. All further references to this edition will be given in the text. Auerbach describes himself as having been "guided by a few motifs which I have worked out *gradually and without a specific purpose*, and in trying them out on a series of texts . . . for I am convinced that these basic motifs in the history of the representation of reality—provided I have seen them correctly—must be demonstrable in *any random* realistic text" (548, emphasis added). He seems to rely on the type of epistemic objectivity that Daston and Galison describe as "trained judgment," or the use of "active and highly trained judgment" (in this case, in the discipline of philology) in order to "see correctly" images that were produced without subjective intervention, guidance, or plan (Daston and Galison, *Objectivity*, 319). In this case, however, the only optical instrument used is close reading.

Note that Auerbach seems to rely on random selection even within his texts. The passages reproduced nearly all come from the middle of their respective texts, similar to Scott's selection of that dead-end scene in *Emma*: the tenth canto of the *Inferno*; a scene from volume 2, chapter 14 of Stendhal's *The Red and the Black*; or a snippet of unimportant conversation between Prince Hal and the very minor character Poins in act 2, scene 2 of *Henry IV, part 2*, to take a few examples.

103. Daston and Galison, *Objectivity*, 319.

104. Here, he describes himself at work: "As I open Rostovtzeff's work to check the quotation above, my eyes fall on this sentence: 'The question, however, arises, How are we to account for the existence of comparatively large numbers of proletarians in Italy?'" (39). Or again, "I open a volume of Shakespeare at random and come upon Macbeth, act 3, scene 6" (325). There seem to be two exceptions: he chooses the "climax" of La Sale's story in chapter 10, and quotes the opening of Schiller's play, *Luise Millerin*.

105. Auerbach's discussion of Dante addresses explicitly the problem of knowing how things turned out. In the passage from the *Inferno* that he quotes, the characters "know the past and the future in the passing of time on earth," but "they are passionately interested in the present state of things on earth, which is hidden from them" (193)—a conundrum that seems to fascinate Auerbach. This is also the only part of the book where he quotes Hegel's *Aesthetics*, borrowing the latter's phrase "changeless existence" from "one of the most beautiful passages ever written on Dante" (191). It's also fascinating that the next event in the Abraham story is withheld from the first chapter of *Mimesis*. Auerbach delays to mention it until *this* chapter, when he quotes the Vulgate: "When Abraham takes the knife to sacrifice his son Isaac, we read: et ecce Angelus Domini de caelo clamavit" (180).

106. See Kent Puckett, *Narrative Theory: A Critical Introduction* (Cambridge: Cambridge University Press, 2016).

107. Foucault, *The Order of Things*, 236–37.

108. Adela Pinch makes this point in her article "Reality Sensing in Elizabeth Gaskell; or, Half-Mended Stockings," *ELH* 83, no. 3 (2016): 821–37, 821–22.

109. It furthers the breakdown of that class-based hierarchy of genres that has been described more recently by Jacques Rancière as the regime of representation. See Jacques Rancière, *The Politics of Aesthetics*, trans. Gabriel Rockhill (London: Continuum, 2004); Jacques Rancière, *Mute Speech*, trans. James Swenson (New York: Columbia University Press, 2011).

110. This may appear to refer to a romantic/modernist technique of symbolization, but Auerbach praises the eighteenth-century French memoirist Saint-Simon for doing the same thing in his treatment of anecdotes: "He alone knows how to use the random and idiosyncratic, the unselected . . . as points of departure for sudden descents into the depths of human existence" (432).

111. See Tzvetan Todorov, "Narrative Transformations," in *The Poetics of Prose*, trans. Richard Howard (Ithaca: Cornell University Press, 1977), 218–33.

112. Virginia Woolf, *To the Lighthouse* (New York: Harcourt, 2005), 211. All further references to this edition will be given in the text.

113. For helpful readings of this chapter, see David Sherman, "A Plot Unraveling into Ethics: Woolf, Levinas, and 'Time Passes,'" *Woolf Studies Annual* 13: 159–79, as well as Stefanie Heine, "Forces of Unworking in 'Time Passes,'" *Textual Cultures* 12, no. 1 (2019): 120–36.

114. For a reading of this passage that connects narratology and phenomenology, see Marco Caraciollo, "Leaping into Space: The Two Aesthetics of *To the Lighthouse*," *Poetics Today* 31, no. 2 (2010): 251–84.

115. On metalepsis, see Freedgood, *Worlds Enough.*

116. For a different reading that finds in such moments something closer to a Deleuzean assemblage, see Derek Ryan, *Virginia Woolf and the Materiality of Theory: Sex, Animal, Life* (Edinburgh: Edinburgh University Press, 2013).

117. On the function of architectural settings and metaphors in the realist novel, see Anna Kornbluh, *The Order of Forms* (Chicago: University of Chicago Press, 2019). My understanding of the significance of setting is somewhat different, though ultimately not incompatible with the argument made by Fredric Jameson, "The Realist Floorplan," in *On Signs*, ed. Marshall Blonsky (Baltimore: Johns Hopkins University Press, 1985), 373–83.

118. Jacques Rancière, "The Thread of the Novel," *Novel: A Forum on Fiction* 47, no. 2 (2014): 196–209, 204.

119. Peter Hanns Reill, *Vitalizing Nature in the Enlightenment* (Berkeley: University of California Press, 2005), 5. Reill argues that this began to shift around the 1740s.

120. A similar commitment to the dimension of plot, though a different set of coordinates, is found in Patricia Spacks, *Desire and Truth* (Chicago: University of Chicago Press, 1990).

121. Two important and relevant accounts of the Victorian novel's relation to science can be found in Gillian Beer, *Darwin's Plots* (Cambridge: Cambridge University Press, 2000); and Levine, *Dying to Know.* I am particularly indebted to Beer's account of the shared discourse of science and literary narrative.

122. Walter Scott, "Methods of Construction," in Williams, *Sir Walter Scott on Novelists and Fiction*, 439–46, 445. Scott is discussing the unpopularity of his own novel, *The Monastery.*

Chapter 2

1. Nancy Armstrong, *How Novels Think* (New York: Columbia University Press, 2005), 5–6.

2. Viktor Shklovsky, "Sterne's *Tristram Shandy*: Stylistic Commentary," in *Russian Formalist Criticism: Four Essays*, ed. Lee T. Lemon and Marion J. Reis (Lincoln: University of Nebraska Press, 1965), 25–60, 57.

3. Laurence Sterne, *The Life and Opinions of Tristram Shandy, Gentleman*, ed. Melvyn New and Joan New (London: Penguin, 1997), 339.

4. Ibid., 8.

5. Quoted in Jonathan Culler, "Defining Narrative Units," in *Style and Structure in Literature*, ed. Roger Fowler (Ithaca: Cornell University Press, 1975), 123–42, 129.

6. Ibid., 139. But Culler also ends this essay with Shklovsky's concept of the "illusory ending," in which Shklovsky argues that "'it is descriptions of nature or of the weather that furnish material for these illusory endings,'" and Culler agrees, though for aesthetic reasons (141).

7. Jacques Rancière, "The Thread of the Novel," *Novel: A Forum on Fiction* 47, no. 2 (2014): 196–209, 204, 198.

8. Ibid., 207.

9. G. W. F. Hegel, *Aesthetics*, trans. T. M. Knox, 2 vols. (Oxford: Clarendon, 1975), 2:1064.

10. Though for a fascinating account of corpuscular science as the study of "imperceptible causality" (12), see Helen Thompson, *Fictional Matter* (Philadelphia: University of Pennsylvania Press, 2017).

11. See, for example, Amy King, *Bloom: The Botanical Vernacular in the English Novel* (New York: Oxford University Press, 2003); Denise Gigante, *Life: Organic Form and Romanticism* (New Haven: Yale University Press, 2009); Devin Griffiths, *The Age of Analogy* (Baltimore: Johns

Hopkins University Press, 2016); Dahlia Porter, *Science, Form, and the Problem of Induction* (Cambridge: Cambridge University Press, 2018). A valuable study of mechanical philosophy in relation to the novel can be found in Joseph Drury, *Novel Machines* (New York: Oxford University Press, 2017). While indebted to them, my argument will differ from those found in the studies by John Bender, *Ends of Enlightenment* (Stanford: Stanford University Press, 2012), Thompson, *Fictional Matter*, and Drury, *Novel Machines*, in ways that will emerge over the course of this chapter. On the overall shape and evolution of natural philosophy, see the essays in Peter Harrison, Ronald L. Numbers, and Michael H. Shank, eds., *Wrestling with Nature* (Chicago: University of Chicago Press, 2011). John Heilbron argues that 1770 marks an "inflection point" in the development and application of natural philosophy (John L. Heilbron, "Natural Philosophy," in Harrison, Numbers, and Shank, 173–99, 174).

12. Steven Shapin, *The Scientific Revolution* (Chicago: University of Chicago Press, 1996), 37.

13. Peter Dear, *Discipline and Experience* (Chicago: University of Chicago Press, 1995). In *Ends of Enlightenment*, John Bender has also argued that "the early novel figures in specific ways in the discursive network now called the scientific revolution" (30). Bender's valuable and engaging study explores the complex links between the novel's stance and the varied protocols and practices of natural philosophy, between fictionality, hypothesis-making, and experimentation (42–49).

14. On the development and significance of such protocols, see Steven Shapin and Simon Schaffer, *Leviathan and the Air-Pump* (Princeton: Princeton University Press, 1985).

15. Though see Margaret Jacob, *The Newtonians and the English Revolution* (Ithaca: Cornell University Press, 1976).

16. See Jacob, *Newtonians*; Margaret Jacob and Larry Stewart, *Practical Matter* (Cambridge, MA: Harvard University Press, 2004).

17. Domenico Bertoloni Meli, *Thinking with Objects* (Baltimore: Johns Hopkins University Press, 2006). I use the strange preposition "around" here in order to suggest how this type of movement from concrete to abstract could preempt the ways in which objectivity would become more of an epistemic concern.

18. Ian Watt, *The Rise of the Novel* (Berkeley: University of California Press, 1957), 15.

19. In some ways, Watt's study can be assimilated to the "two cultures" thesis, as influentially articulated by C. P. Snow in a lecture given two years after the publication of Watt's book. Complaining of the "total incomprehension of science" on the part of "literary intellectuals," Snow claims that if he had asked a group of the latter, "What do you mean by mass, or acceleration . . . the scientific equivalent of saying, Can you read?—not more than one in ten of the highly educated would have felt that I was speaking the same language." C. P. Snow, *The Two Cultures* (Cambridge: Cambridge University Press, 1998), 11, 15. Michael McKeon's *The Origins of the English Novel* (Baltimore: Johns Hopkins University Press, 1987) offers an account equally rich in theory and detail, with attention to the Royal Society and its protocols for studying and transcribing experience.

20. See, for example, the account of the "new experimentalists" in Deborah Mayo, *Error and the Growth of Experimental Knowledge* (Chicago: University of Chicago Press, 2014).

21. Gillian Beer notes in *Darwin's Plots* (Cambridge: Cambridge University Press, 2000) that this remained the case well into the Victorian era: there was "nothing hermetic or exclusive in the writing of Lyell or Darwin . . . [it was] readily available to readers without a scientific training" (4). On the use of shared tropes, see Tita Chico's excellent study, *The Experimental Imagination* (Stanford: Stanford University Press, 2018). See also Courtney Weiss Smith,

Empiricist Devotions (Charlottesville: University of Virginia Press, 2016), on the dissemination of Newtonian science, as well as Jacob and Stewart, *Practical Matter*.

22. Michael Hunter, *Science and Society in Restoration England* (Cambridge: Cambridge University Press, 1981), 84–85, 191–92. See also, for example, Larry Stewart, "The Laboratory, the Workshop, and the Theatre of Experiment," in *Science and Spectacle in the European Enlightenment*, ed. Bernadette Bensaude-Vincent and Christine Blondel (Hampshire, England: Ashgate, 2008), 11–24, as well as the essays in Christine Lehleiter, ed., *Fact and Fiction: Literary and Scientific Cultures in Germany and Britain* (Toronto: University of Toronto Press, 2016).

23. Crucially, as Weiss Smith's *Empiricist Devotions*, and Heilbron's "Natural Philosophy," show, Newtonian natural philosophy was largely dependent on these popularizing explanations, as the math involved in Newton's published works, even when translated into English, exceeded the capacity of most people to comprehend.

24. Pamela Smith, in *The Body of the Artisan* (Chicago: University of Chicago Press, 2004), argues that the "epistemology of handwork" developed by artisans contributed to the development of science in this period (28).

25. John Desaguliers, *A Course of Experimental Philosophy*, 2 vols. (London, 1734), 1:3–5.

26. See Michael Marrinan and John Bender, *The Culture of Diagram* (Stanford: Stanford University Press, 2010).

27. Jonathan Israel, *Radical Enlightenment* (Oxford: Oxford University Press, 2001), 151. A similar interpretation is found in a recent study by Evander Agazzi, who lays more stress on the "anti-essentialist and anti-substantialist features" that characterized the new modern science (*Scientific Objectivity and Its Contexts* [Cham, Switzerland: Springer, 2014], 3).

28. Shapin, *Scientific Revolution*, 5.

29. Adam Smith, *Essays on Philosophical Subjects* (London, 1795), 59, 104. On Smith as an Enlightenment system maker, see Clifford Siskin, "Mediated Enlightenment: The System of the World," in *This Is Enlightenment*, ed. Clifford Siskin and William Warner (Chicago: University of Chicago Press, 2010), 164–72.

30. Adam Smith, *Essays*, 59.

31. Shapin, *Scientific Revolution*, 29, 37.

32. Israel, *Radical Enlightenment*, 17.

33. Willem Jacob s'Gravesande, *Mathematical Elements of Natural Philosophy Confirmed by Experiments, or an Introduction to Newton's Philosophy*, trans. J. T. Desaguliers, 2nd ed. (London: Taylor, 1721), 26.

34. Robert Boyle, "Of the Excellency and Grounds of the Corpuscular or Mechanical Philosophy," in *Nature and Nature's Laws*, ed. Marie Boas Hall (New York: Walker, 1970), 311–23, 313. For a detailed and nuanced study of how Boyle's corpuscular philosophy allows for invisible, imperceptible forces, see Thompson, *Fictional Matter*.

35. Boyle, "Excellency and Grounds," 314.

36. Ibid., 312.

37. Desaguliers, *Course of Experimental Philosophy*, 1:22–23.

38. The genre of "it-narratives" often traces such journeys; see, for example, the essays in Mark Blackwell, ed., *The Secret Life of Things* (Lewisburg, PA: Bucknell University Press, 2007).

39. Adam Smith, *Essays*, 76. See Peter Hanns Reill, *Vitalizing Nature in the Enlightenment* (Berkeley: University of California Press, 2005).

40. Jacob, *Newtonians*, 65.

41. Israel, *Radical Enlightenment*, 11.

42. Oliver Goldsmith, *A Survey of Experimental Philosophy, Considered in Its Present State of Improvement* (London, 1776), 106–7.
43. Adam Smith, *Essays*, 76.
44. Desaguliers, *Course of Experimental Philosophy*, 1:17.
45. On the print/book form of the *Principia* and its popularization, see Laura Miller, *Popular Newtonianism* (Charlottesville: University of Virginia Press, 2018); see also Jacob and Stewart, *Practical Matter*.
46. See, for example, Drury, *Novel Machines*; Jessica Riskin, *The Restless Clock* (Chicago: University of Chicago Press, 2016).
47. Isaac Newton, *Principia*, trans. Andrew Motte, 3 vols. (London: Symond, 1803), 1:14. Emphasis added.
48. Ibid., 15.
49. Benjamin Martin, *A plain and familiar introduction to the Newtonian philosophy* (London, 1754), 149.
50. Newton, *Principia*, 2:159. On the significance of the system to the eighteenth century, see Siskin, "Mediated Enlightenment."
51. Adam Smith, *Essays*, 93. Smith attempts to find the laws of moral phenomena in Adam Smith, *The Theory of Moral Sentiments*, ed. D. D. Raphael and A. L. Macfie (Indianapolis: Liberty Fund, 1982), putting the principle of sympathy in the place of gravity.
52. Desaguliers, *Course of Experimental Philosophy*, 1:340.
53. Newton, *Principia*, 1:14.
54. The conventional narratological model of plot often begins with a situation of stasis or equilibrium which is disrupted by the force of some event. For a useful and comprehensive discussion, see Brian Richardson, *A Poetics of Plot for the Twenty-First Century* (Columbus: Ohio State University Press, 2019), especially 59–82.
55. See Zvi Biener, "Newton's Regulae Philosophandi," in *The Oxford Handbook of Newton*, ed. Eric Schliesser and Chris Smeenk (Oxford: Oxford University Press, 2017), 1–25.
56. Newton, *Principia*, 2:160.
57. Ibid., 1:3–4.
58. Ibid., 2:310–11.
59. Israel, *Radical Enlightenment*, 518.
60. Jacob, *Newtonians*, 29, 52, 60. See also Weiss Smith, *Empiricist Devotions*.
61. Newton, *Principia*, 2:160.
62. Ibid., 2:313–14.
63. s'Gravesande, *Mathematical Elements*, 2.
64. Ibid., 18.
65. Desaguliers, *Course of Experimental Philosophy*, 1:6–7. On theories of ethereal fluids, see Patricia Fara, *Sympathetic Attractions* (Princeton: Princeton University Press, 1996), 182–83.
66. Bernard Fontenelle, *The Life of Sir Isaac Newton* (London, 1728), 19.
67. Agazzi, *Scientific Objectivity*, 4.
68. Shapin, *Scientific Revolution*, 101.
69. Shapin and Schaffer, *Leviathan and the Air-Pump*, 19–20.
70. Quoted in Agazzi, *Scientific Objectivity*, 3n3.
71. Weiss Smith, *Empiricist Devotions*, 69–105.
72. s'Gravesande, *Mathematical Elements*, 2.
73. Ibid., 3.

74. Boyle, "Excellency and Grounds," 316 (emphasis added). On actor-network theory, see Bruno Latour, *Reassembling the Social* (Oxford: Oxford University Press, 2005).

75. Boyle, "Excellency," 317.

76. Ibid., 318.

77. Ibid., 321, emphasis added.

78. Goldsmith, *Survey of Experimental Philosophy*, 165.

79. Ibid., 176.

80. Desaguliers, *Course of Experimental Philosophy*, 1:42.

81. Dear, *Discipline and Experience*, 246.

82. Ibid., 20–21.

83. Ibid., 63–64.

84. Ibid., 69–70.

85. Ibid., 77.

86. Ibid., 126, original emphasis.

87. Ibid., 228–29, original emphasis.

88. s'Gravesande, *Mathematical Elements*, 13.

89. Goldsmith, *Survey of Experimental Philosophy*, 31–32.

90. Shapin and Schaffer, *Leviathan and the Air-Pump*, 60–65.

91. Goldsmith, *Survey of Experimental Philosophy*, 46–49.

92. Ibid., 23–25.

93. On virtual witnessing, see also Marrinan and Bender, *Culture of Diagram*.

94. Desaguliers, *Course of Experimental Philosophy*, 1:21, emphasis added.

95. Thompson, *Fictional Matter*, 20.

96. Ibid., 285.

97. Harris was a public science lecturer, encyclopedist, and Boyle lecturer; see Jacob and Stewart, *Practical Matter*, 64–76.

98. John Harris, *Astronomical Dialogues Between a Gentleman and a Lady: Wherein the Doctrine of the Spheres, Uses of the Globes, and the Elements of Astronomy and Geography are Explain'd, In a Pleasant, Easy, and Familiar Way* (London, 1719), 4.

99. Ibid., 91.

100. Ibid., 90, original emphasis.

101. Martin, *Plain and familiar introduction*, 139, original emphasis.

102. Ibid., 138–39, original emphasis.

103. Ibid., 142.

104. Bertoloni Meli, *Thinking with Objects*, 6, 225.

105. Ibid., 5. This might be compared to Bruno Latour's description of "operators," a device "which belongs to matter at one end, to form at the other. The operators are linked in a series that passes across the difference between things and words . . . the earth becomes a cardboard cube, words become paper, colors become numbers, and so on." (Bruno Latour, *Pandora's Hope* [Cambridge, MA: Harvard University Press, 1999], 69).

106. Judith Hawley notes, "Mechanical philosophy sometimes seems to govern the Shandy Household" in her discussion of Sterne's relation to learned wit and natural philosophy, in "Tristram Shandy, Learned Wit, and Enlightenment Knowledge," in *The Cambridge Companion to Laurence Sterne*, ed. Thomas Keymer (Cambridge: Cambridge University Press, 2009), 34–48, 35.

107. See, for example, Drury, *Novel Machines*, 108–42.

108. On Sterne's indebtedness to the tradition of learned wit, see Thomas Keymer, *Sterne, the Moderns, and the Novel* (Oxford: Oxford University Press, 2002); for a reading of Sterne in relation to the sentimental tradition in moral philosophy and the novel, see James Chandler, "The Novelty of Laurence Sterne," in *English and British Fiction, 1750–1820*, ed. Peter Garside and Karen O'Brien (Oxford: Oxford University Press, 2015), 109–28.

109. Shklovsky, "Sterne's *Tristram Shandy*," 27; on the maximum-machine, see Alan Morton, "Concepts of Power," *British Journal of the History of Science* 28, no. 1 (1995): 63–78, 72.

110. See, for example, Deidre Lynch, "The Shandean Lifetime Reading Plan," in *The Work of Genre: Selected Essays from the English Institute*, ed. Robyn Warhol-Down (Cambridge, MA: English Institute in collaboration with the American Council of Learned Societies, 2011). Weiss Smith, *Empiricist Devotions*, discusses the clock metaphor in detail, including Leibniz's criticism that, "according to their Doctrine, God Almighty wants to wind up his Watch from Time to Time: Otherwise it would cease to move" (quoted, 72).

111. Compare Bruno Latour's account of soil sampling in the Amazon in *Pandora's Hope*, 24–79.

112. See Jessica Riskin, "Eighteenth-Century Wetware," *Representations* 83 (2003): 97–125.

113. On the spring of the air as interpreted by Boyle, see Thompson, *Fictional Matter*, 12–13.

114. Shapin, *Scientific Revolution*, 44.

Chapter 3

1. Homer Obed Brown, *Institutions of the English Novel from Defoe to Scott* (Philadelphia: University of Pennsylvania Press, 1997), 177.

2. Ibid., 177–78.

3. Ruth Yeazell, *Art of the Everyday* (Princeton: Princeton University Press, 2008), 1–4.

4. Hippolyte Taine, *Histoire de la littérature anglaise* (Paris, 1863), quoted in Yeazell, 5.

5. Anna Laetitia Barbauld, "On the Origin and Progress of Novel-Writing," *The British Novelists*, vol. 1 (London: Rivington, 1820), 15.

6. Brown, *Institutions of the English Novel*, 197.

7. Daniel Defoe, *The Storm*, ed. Richard Hamblyn (New York: Penguin, 2005), 59.

8. On the relevant context of contemporary journalism, especially the literature of wonder as well as more didactic "Providence books," see J. Paul Hunter, *Before Novels* (New York: Norton, 1990), 195–224.

9. *Aristotle's "Poetics,"* trans. James Hutton (New York: Norton, 1982), 56.

10. Erich Auerbach, *Mimesis* (Princeton: Princeton University Press, 1974), 548.

11. Daniel Defoe, *Moll Flanders*, ed. David Blewett (London: Penguin, 1989), n.p.

12. Daniel Defoe, *Robinson Crusoe*, ed. John Richetti (London: Penguin, 2001), 16. Further references to this edition will be given in the text.

13. I hope to offer an alternative way of understanding Defoe's realism, which has been much discussed; see, for example, Maximillian Novak, *Realism, Myth, and History in Defoe's Fiction* (Lincoln: University of Nebraska Press, 1983) and, more recently, Franco Moretti, *The Bourgeois* (New York: Verso, 2013).

14. Willem Jacob s'Gravesande, *Mathematical Elements of Natural Philosophy Confirmed by Experiments, or an Introduction to Newton's Philosophy*, trans. J. T. Desaguliers, 2nd ed. (London: Taylor, 1721), 2.

15. Defoe, *The Storm*, 160.

16. Fredric Jameson, *The Antinomies of Realism* (New York: Verso, 2013), 211.

17. Moretti, *The Bourgeois*, 51–66.

18. On metalepsis and realism, see Elaine Freedgood, *Worlds Enough* (Princeton: Princeton University Press, 2019).

19. Roland Barthes, "The Reality Effect," in *The Rustle of Language*, trans. Richard Howard (Berkeley: University of California Press, 1986), 141–48.

20. Carl Fisher, "Innovation and Imitation in the Eighteenth-Century Robinsonade," in *The Cambridge Companion to Robinson Crusoe*, ed. John Richetti (Cambridge: Cambridge University Press, 2018), 99–111. In addition to its European and transatlantic reach, Defoe's novel was also translated into Chinese and Japanese, and had a particularly profound influence in China. It continues to be adapted and assigned to schoolchildren; see Haifeng Hui, "Literary Classics, Consumer Culture, and Chinese Children's Educational Book Market: Material Parameters and Thematic Adaptations in New Curricular Editions of Robinson Crusoe," CLCWeb 19, 3 (2017); see also in Maxine Hong Kingston's *China Men* (New York: Knopf, 1980), the story of "Lo Bun Sun," as well as her use of it as a generalizable paradigm for the male Chinese immigrants to America who provide the subject of this book.

21. As a sample: Ian Watt, *The Rise of the Novel* (Berkeley: University of California Press, 1957); Michael McKeon, *Origins of the English Novel* (Baltimore: Johns Hopkins University Press, 1987); J. Paul Hunter, *The Reluctant Pilgrim* (Baltimore: Johns Hopkins University Press, 1966); John Richetti, *Defoe's Narratives* (Oxford: Clarendon, 1975); Dennis Todd, "Robinson Crusoe and Colonialism," in *Cambridge Companion to Robinson Crusoe*, 142–56; J. M. Coetzee, *Foe* (New York: Penguin, 1987); Margaret Cohen, *The Novel and the Sea* (Princeton: Princeton University Press, 2010).

22. On narrativity and narratability, see Gerald Prince, "Narrativehood, Narrativeness, Narrativity, Narratability," in *Theorizing Narrativity*, ed. John Pier and José Ángel García Landa (Berlin: De Gruyter, 2008), 19–27.

23. See Jonathan Kramnick, *Actions and Objects from Hobbes to Richardson* (Stanford: Stanford University Press, 2010).

24. Ilse Vickers, *Defoe and the New Sciences* (Cambridge: Cambridge University Press, 1996), 31.

25. See Vickers, *Defoe and the New Sciences*, 55–65.

26. Daniel Defoe, *A general history of discoveries and improvements* (London, 1725–26), 251.

27. Ibid., 252, 254.

28. Ibid., 254.

29. Ibid., 257–58.

30. Robert Boyle, "Of the Excellency and Grounds of the Corpuscular or Mechanical Philosophy," in *Nature and Nature's Laws*, ed. Marie Boas Hall (New York: Walker, 1970), 316, emphasis added.

31. Ibid.

32. s'Gravesande, *Mathematical Elements*: "We do not exclude out of the number of natural phaenomena, those which happen in our bodies by our will; for they are produced by the motion of our muscles, and their action depends upon another motion: in these, there is only that motion which arises from the immediate action of the mind, and is entirely unknown to us, which is not a natural phaenomenon" (2).

33. On the literature of wonder, as well as literature and wonder, see J. Paul Hunter, *Before Novels* (New York: Norton, 1990); and Sarah Kareem, *Eighteenth-Century Fiction and the*

Reinvention of Wonder (Oxford: Oxford University Press, 2014), which argues that "eighteenth-century fiction cultivates wonder as a response to the ordinary in a manner consistent with the Humean critique of induction" (77).

34. Defoe, *The Storm*, 13, original emphasis.

35. Ibid., 14.

36. Ibid., 14–15, emphasis added.

37. Ibid., 11.

38. Ibid., 14, emphasis added.

39. Defoe argues that the natural philosopher's "Search after Causes" responds to a challenge or puzzle set by divine power for humans to solve: "To search after what our Maker has not hid, only cover'd with a thin Veil of Natural Obscurity . . . seems to be justified by the very Nature of the thing." A mere sense of wonder is a sign of the failure of science: "In Nature the Philosopher's Business is not to look through Nature, and come to the vast open Field of Infinite Power. . . . The Christian begins just where the Philosopher ends; and when the Enquirer turns his Eyes up to Heaven, Farewel Philosopher; 'tis a Sign he can make nothing of it here" (14).

40. John Desaguliers, *A Course of Experimental Philosophy*, vol. 1 (London, 1734), 21.

41. I depart from John Richetti's account of "a material world that is random, without purpose or pattern," in John Richetti, "Defoe as Narrative Innovator," in *The Cambridge Companion to Daniel Defoe*, ed. John Richetti (Cambridge: Cambridge University Press, 2008), 121–38, 124–25. The natural world is regulated by its own laws, even if those laws are not fully known, Defoe suggests.

42. Daniel Defoe, *The Farther Adventures of Robinson Crusoe* (London: Taylor, 1719), 10–11, emphasis added.

43. Defoe, *Discoveries and Improvements*, 2. Many have emphasized the practical orientation of Baconian science; see, for example, Vickers, *Defoe and the New Sciences*.

44. Defoe, *The Storm*, 13.

45. Planting is actually one of the most important actions in the novel, besides forming one of the three "great branches" of useful arts. "Correspondence of People . . . necessarily begat Trade, Trade begat Navigation, Navigation by making Discoveries, begat Plantation, and remote Plantations again encreas'd Correspondence" (Defoe, *Discoveries and Improvements*, 79). When departing the island, Crusoe leaves behind for the Spaniards a "long Paper of Directions . . . the particular Methods which I took for managing every Part of my Life there, the Way how I baked my Bread, bred up tame Goats, and planted my Corn . . . in a word, every Thing I did, all this being written down" (Defoe, *Farther Adventures*, 46); the method of planting is transmitted in turn even to the "savages" whom the Europeans eventually allow to inhabit one corner of the island.

46. On the canoe as a transcultural object, see Peter Walmsley, "Robinson Crusoe's Canoes," *Eighteenth Century Life* 43, no. 1 (2019): 1–23. Walmsley points out that the too-heavy canoe does not play a part in the plot; thus, "The problem of the savage is negotiated in the realm of work" (16).

47. On mistakes and problem-solving as part of the mariner's craft, see Cohen, *The Novel and the Sea*, 7–8.

48. Amélie Oksenberg Rorty, "The Psychology of Aristotelian Tragedy," in *Essays on Aristotelian Poetics*, ed. Amélie Oksenberg Rorty (Princeton: Princeton University Press, 1992), 1–22, 3.

49. Ibid., 5.

50. See Jean-Pierre Vernant, "Myth and Tragedy," in Rorty, *Essays on Aristotelian Poetics*, 33–50.

51. Ibid., 41.

52. Terence Cave, *Recognitions: A Study in Poetics* (Oxford: Clarendon Press, 1990).

53. Building on Walmsley's argument in "Robinson Crusoe's Canoes," we might say that the "problem of the savage" is also worked out in the realm of recognition.

54. Cave, *Recognitions*, 1.

55. See Coetzee's *Foe*, in which the endless activity is terrace-building.

56. Moretti, *The Bourgeois*, 71, original emphasis. His reading of Robinson Crusoe also points out both the "solidity" of this world, and its curious openness toward the future, which Moretti aligns with capitalism, modernity, and the commitment to progress. Moretti aligns Defoe's style with Hegel's notion of the "prose of the world," but viewed through the lens of Weber and the young Lukács.

57. Moretti, *The Bourgeois*, 75, original emphasis.

58. Desaguliers, *Course of Experimental Philosophy*, 1:42.

59. Discussed in the previous chapter.

60. Desaguliers, *Course of Experimental Philosophy*, 1:3–5.

61. On the description of things in this novel, see Cynthia Wall, *The Prose of Things* (Chicago: University of Chicago Press, 2006).

62. See Srinivas Aravamudan, "Defoe, Commerce, and Empire," in *The Cambridge Companion to Daniel Defoe*, ed. John Richetti (Cambridge: Cambridge University Press, 2008), 45–63.

63. Jacques Rancière, *Dissensus*, ed. and trans. Steven Corcoran (London: Continuum, 2010), 36.

64. G. W. F. Hegel, *Aesthetics*, trans. T. M. Knox, 2 vols. (Oxford: Clarendon, 1975), 2:1051.

65. Ibid., 2:1066, 2:1070.

66. Ibid., 2:1071.

Chapter 4

1. The laws in this case are not social or juridical; they claim to describe the physical world but are produced through human investigation. See the remarkable argument about the former in Sandra Macpherson, *Harm's Way* (Baltimore: Johns Hopkins University Press, 2010). For a thoughtful analysis of setting into the concepts of "ground" and "milieu" in modernist fiction, see Hannah Freed-Thall and Dora Zhang, "Modernist Setting," *Modernism/Modernity* 3, no. 1 (2018), https://doi.org/10.26597/mod.0042.

2. For a rich discussion of typicality and character in the realist novel, see Catherine Gallagher, "George Eliot, Immanent Victorian," *Representations* 90, no. 1 (2005): 61–74. Gallagher similarly argues that "Novelists took the abstract entity, the species or type, to be the given, the thing-in-the-world referent grounding the form" (62).

3. See, for example, Ann Blair, *The Theater of Nature* (Princeton: Princeton University Press, 1997).

4. On the affordances offered by forms, see Caroline Levine, *Forms* (Princeton: Princeton University Press, 2015).

5. On the importance of analogy, see Devin Griffiths, *The Age of Analogy* (Baltimore: Johns Hopkins University Press, 2016).

6. For an influential discussion of the emergence of fictionality in the eighteenth-century context, see Catherine Gallagher, "The Rise of Fictionality," in *The Novel*, ed. Franco Moretti, vol. 1 (Princeton: Princeton University Press, 2006), 336–63.

7. William Hazlitt, *Lectures on the English Comic Writers* (London: Oxford University Press, 1907), 148.

8. Ibid., emphasis added.

9. On the relations between the novel and natural history, see Amy King, "Dilatory Description and the Pleasures of Accumulation," in *Narrative Middles*, ed. Caroline Levine and Mario Ortiz-Robles (Columbus: Ohio State University Press, 2011), 161–94, as well as Robert James Merrett, "Natural History and the Eighteenth-Century Novel," *Eighteenth-Century Studies* 25, no. 2 (1991–92): 145–70. See also Amy King, *Bloom: The Botanical Vernacular in the English Novel* (New York: Oxford University Press, 2003), on links between botany and the marriage plot.

10. See Michael McKeon, *Origins of the English Novel* (Baltimore: Johns Hopkins University Press, 1987), 65–73 in particular.

11. Such activities were also bound up from the start with European colonization and the exploitation of other races and women; see, for example, Mary Louise Pratt, *Imperial Eyes* (New York: Routledge, 1992); Londa Schiebinger, *Nature's Body* (Boston: Beacon, 1993); Alan Bewell, *Natures in Translation* (Baltimore: Johns Hopkins University Press, 2017).

12. Harold Cook, "Physicians and Natural History," in *Cultures of Natural History*, ed. Nicholas Jardine, J. A. Secord, and E. C. Spary (Cambridge: Cambridge University Press, 1996), 91–105, 104.

13. *Philosophical Transactions of the Royal Society* 42 (1742–43): 218–20.

14. Paul Farber, *Finding Order in Nature* (Baltimore: Johns Hopkins University Press, 2000), 77.

15. Katie Whitaker, "The Culture of Curiosity," in Jardine, Secord, and Spary, *Cultures of Natural History*, 75–90, 87.

16. Ibid., 87.

17. Martin Rudwick, "Minerals, Strata, and Fossils," in Jardine, Secord, and Spary, *Cultures of Natural History*, 266–86, 269.

18. Michel Foucault, *The Order of Things* (New York: Random House, 1970), 144.

19. Ibid., 50.

20. Ibid., 55.

21. Ibid., 133–34.

22. Ibid., 67.

23. Ibid., 73.

24. See Clifford Siskin, "Mediated Enlightenment: The System of the World," in *This Is Enlightenment*, ed. Clifford Siskin and William Warner (Chicago: University of Chicago Press, 2010), as well as the remarkable study of the *Encyclopédie* in Michael Marrinan and John Bender, *The Culture of Diagram* (Stanford: Stanford University Press, 2010).

25. Charles Withers, "Geography, Natural History and the Eighteenth-Century Enlightenment: Putting the World in Place," *History Workshop Journal* 39 (1995): 136–63, 139.

26. Quoted in Withers, "Geography, Natural History," 141.

27. See William Hogarth, *Hogarth's Graphic Works*, ed. Ronald Paulson (New Haven: Yale University Press, 1965). For an important reading of Hogarth in the context of experimental

philosophy, see John Bender, *Ends of Enlightenment* (Stanford: Stanford University Press, 2012), and the brilliant revision of it offered by Helen Thompson in *Fictional Matter* (Philadelphia: University of Pennsylvania Press, 2017), 16–21.

28. On the type of "scene shift" involving metalepsis, see Monika Fludernik, "Scene Shift, Metalepsis, and the Metaleptic Mode," *Style* 37, no. 4 (2003): 382–400.

29. The chronotope is the term invented by Mikhail Bakhtin in "Forms of Time and of the Chronotope in the Novel," in *The Dialogic Imagination*, ed. Michael Holquist, trans. Caryl Emerson and Michael Holquist (Austin: University of Texas Press, 1981), 84–258.

30. Foucault, *The Order of Things*, 268.

31. David Hume, *A Treatise of Human Nature*, ed. L. A. Selby-Bigge (Oxford: Clarendon, 1989), 104–5. Here, I build on the argument made by Ian Duncan about the centrality of David Hume to the rising status of fictional narrative in and after Walter Scott's time. In Ian Duncan, *Scott's Shadow* (Princeton: Princeton University Press, 2007), Duncan argues that "Hume's case, that all representation is a fiction . . . since all experience is mediated through the imagination . . . [delivers fiction] from the sentence of inauthenticity, of categorical opposition to reality" (133).

32. Samuel Richardson, *Clarissa*, ed. Angus Ross (Harmondsworth: Penguin, 1985), 1495, 1498. Further references to this edition will be given in the text.

33. Henry Fielding, *Joseph Andrews*, ed. Homer Goldberg (New York: Norton, 1987), 154. Further references to this edition will be given in the text.

34. *Aristotle's "Poetics,"* trans. James Hutton (New York: Norton, 1982), 71.

35. Ibid., 53.

36. See, for example, Lennard Davis, *Factual Fictions* (New York: Columbia University Press, 1983).

37. Eric Downing, "Οἷον Ψυχή: An Essay on Aristotle's 'Muthos,'" *Classical Antiquity* 3, no. 2 (1984): 164–78, 173.

38. Richard Pulteney, *A general view of the writings of Linnæus. By Richard Pulteney, M. D. & F. R. S.* (London, 1781), 24–25.

39. Ibid., 63.

40. On the significance of Buffon, including his influence on Adam Ferguson, see Peter Hanns Reill, *Vitalizing Nature in the Enlightenment* (Berkeley: University of California Press, 2005).

41. Phillip Sloan, "Kant on the History of Nature: The Ambiguous Heritage of the Critical Philosophy for Natural History," *Studies in History and Philosophy of Biological and Biomedical Sciences* 37 (2006): 627–48, 633.

42. Georges Louis LeClerq, Comte de Buffon, "The 'Initial Discourse' to Buffon's *Histoire naturelle*: The First Complete English Translation," trans. John Lyon, *Journal of the History of Biology* 9, no. 1 (1976): 133–81, 177.

43. Ibid., 176.

44. Ibid., 154.

45. Buffon argues that, depending on which characteristics you select, you could say that a lion is more like a bat than a horse like a dog, that a horse is more like a pig than a dog, etc. (Buffon, "'Initial Discourse,'" 163–64).

46. Ibid., 177.

47. Ibid., 157.

48. Ibid., 150–55.

49. Ibid., 160.

50. Georges Louis Leclerc, Comte de Buffon, *The natural history of animals, vegetables, and minerals; with the theory of the earth in general. Translated from the French of Count de Buffon. Intendant of the Royal Gardens in France; Member of the French Academy, of the Academy of Sciences, and of the Royal Societies of London Berlin, &c. By W. Kenrick, L.L.D. and J. Murdoch*, vol. 1 (London: Bell, [1775]–76), 293.

51. Buffon, "'Initial Discourse,'" 150.

52. Peter Hanns Reill argues, in *Vitalizing Nature*, that Buffon's attitude characterizes the "radical Enlightenment" of the mid-eighteenth century: "Relation, rapport . . . reciprocal interaction replaced aggregation and strict causal relations as defining principles of matter. Identity and noncontradiction were replaced with degrees of relation and similarity" (7).

53. Buffon, *Natural history*, 1:397–98.

54. Buffon, "'Initial Discourse,'" 162.

55. See Lorraine Daston, *Classical Probability in the Enlightenment* (Princeton: Princeton University Press, 1988).

56. Buffon, "'Initial Discourse,'" 174–75, emphasis added.

57. On probability in the novel, see Jesse Molesworth, *Chance and the Eighteenth-Century Novel* (Cambridge: Cambridge University Press, 2010).

58. Buffon, "'Initial Discourse,'" 155.

59. Ibid., 175.

60. Deidre Lynch, *The Economy of Character* (Chicago: University of Chicago Press, 1998), 38.

61. Theophrastus was a natural historian, Aristotle's successor and possible collaborator; he produced works of mineralogy and botany.

62. *Characterism; or, the modern age display'd; being an attempt to expose the pretended virtues of both sexes; with a poetical essay on each character. In two parts. First, of the ladies, second, of the gentlemen* (London, [1750?]).

63. Buffon, *The natural history of animals, vegetables, and minerals; with the theory of the earth in general. Translated from the French of Count de Buffon. Intendant of the Royal Gardens in France; Member of the French Academy, of the Academy of Sciences, and of the Royal Societies of London Berlin, &c. By W. Kenrick, L.L.D. and J. Murdoch*, vol. 2 (London, [1775]–76).

64. *Characterisms*, 20.

65. Buffon, "'Initial Discourse,'" 160.

66. Ibid., 150–52.

67. On caricature and Hogarth, see Lynch, *Economy of Character*, especially 61–70.

68. The canonical approach to Fielding's plots, which dates to a remark by Coleridge, stresses their formal perfection; see the influential essay by R. S. Crane, "The Concept of Plot and the Plot of *Tom Jones*," in *Critics and Criticism*, ed. R. S. Crane (Chicago: University of Chicago Press, 1952), 62–93. But see also John Richetti's reading of the sociohistorical realism that is also vital to this novel's plot in "Formalism and Historicity Reconciled in Henry Fielding's *Tom Jones*," in *Narrative Concepts in the Study of Eighteenth-Century Literature*, ed. Liisa Steinby, Aino Mäkikalli (Amsterdam: Amsterdam University Press, 2017), 79–97.

69. Ian Watt, *The Rise of the Novel* (Berkeley: University of California Press, 1957), 271.

70. It is true, though, that the stories of Wilson and the Man on the Hill, in *Tom Jones*, seem to be hybrids of character-sketch and narrative.

71. Walter Scott, *Waverley*, ed. Claire Lamont (Oxford: Oxford University Press, 1986), 41.

72. Ibid., 91–92.

73. Ibid., 102.

74. Ibid., 338.

75. For the argument that the act of interpretation is central to Clarissa, see William Warner, *Reading Clarissa* (New Haven: Yale University Press, 1979); and Terry Castle, *Clarissa's Ciphers* (Ithaca: Cornell University Press, 1982); here, I focus on one particular type of interpretive practice, loosely defined, that figured importantly in the eighteenth century: the practice of natural history. Others have noted the importance of testing to the plot and its resemblance to experimental practice: see Ruth Perry, *Women, Letters, and the Novel* (New York: AMS, 1980); and Ann Jessie van Sant, *Eighteenth-Century Sensibility and the Novel* (Cambridge: Cambridge University Press, 1993).

76. On the importance of the "tested woman plot" to the Western literary tradition, see Lois Bueler, *Clarissa's Plots* (Newark: University of Delaware Press, 1994). Bueler does not, however, locate the concept of woman in the natural historical context.

77. There is also the symbol of the lily that Clarissa herself chooses with which to adorn her coffin.

78. See Bueler, *Clarissa's Plots*, 22–40.

79. See Hilary Dannenberg, *Coincidence and Counterfactuality* (Lincoln: University of Nebraska Press, 2008); and Catherine Gallagher, *Telling It Like It Wasn't* (Chicago: University of Chicago Press, 2018).

80. See Marrinan and Bender, *The Culture of Diagram.*

Chapter 5

1. Ann Radcliffe, *The Italian*, ed. Frederick Garber (Oxford: Oxford University Press, 1981), 76. All further references will be given in the text.

2. Ann Radcliffe, *The Mysteries of Udolpho*, ed. Bonamy Dobrée (Oxford: Oxford University Press, 1998), 449. All further references will be given in the text.

3. J. T. Desaguliers, *A Course of Experimental Philosophy*, vol. 1 (London, 1734), 3–5.

4. William Lewis, *Commercium philosophico-technicum; or, the philosophical commerce of arts: designed as an attempt to improve arts, trades, and manufactures* (London: Baldwin, 1763–65), 234–35.

5. Lewis, *Commercium philosophico-technicum*, 232–33.

6. Alistair Duncan, *Laws and Order in Eighteenth-Century Chemistry* (Oxford: Clarendon, 1996), 6.

7. Ibid., 10.

8. Ursula Klein and Wolfgang Lefèvre, *Materials in Eighteenth-Century Science* (Cambridge, MA: MIT Press, 2007), 21–22.

9. On the moral power of the "already existent," see Anne-Lise François, *Open Secrets* (Stanford: Stanford University Press, 2008), 247.

10. Lavoisier provided one of Kuhn's models of scientific revolution; see Thomas Kuhn, *The Structure of Scientific Revolutions* (Chicago: University of Chicago Press, 2012). On the chemical studies of Robert Boyle, see Marie Boas, *Robert Boyle and Seventeenth-Century Chemistry* (Cambridge: Cambridge University Press, 1958).

11. Klein and Lefèvre, *Materials in Eighteenth-Century Science*, 21.

12. Ibid., 27.

13. Ibid.

14. Ibid., 20.

15. Ibid., 42.

16. Ibid., 49. See also Victor Boantza, *Matter and Method in the Long Chemical Revolution* (Farnham, Surrey: Ashgate, 2013).

17. Peter Hanns Reill, *Vitalizing Nature in the Enlightenment* (Berkeley: University of California Press, 2005), 73–74.

18. Ibid., 79.

19. Lewis, *Commercium philosophico-technicum*, iv.

20. On the struggle of Lewis and other chemists "to preserve chemical knowledge and set it apart from physical theory," see Boantza, *Long Chemical Revolution*, 10.

21. See Duncan, *Laws and Order in Eighteenth-Century Chemistry*, on the history of affinity as well as various theories.

22. Ibid., 32.

23. Quoted in Duncan, *Laws and Order in Eighteenth-Century Chemistry*, 157.

24. Reill, *Vitalizing Nature*, 84.

25. Jane Marcet, *Conversations on Chemistry: In Which the Elements Are Familiarly Explained and Illustrated by Experiments, in Two Volumes*, 5th ed. (London: Longman, Hurst, 1817), 21–22.

26. Johann Wolfgang von Goethe, *Elective Affinities*, trans. David Constantine (Oxford: Oxford University Press, 1994), 35.

27. Ibid., 34.

28. For an interesting argument about Walter Benjamin's views on character and decision in relation to this novel, see N. K. Leacock, "Character, Silence, and the Novel: Walter Benjamin on Goethe's *Elective Affinities*," *Narrative* 10, no. 3 (2002): 277–306.

29. Marcet, *Conversations on Chemistry*, 1, emphasis added.

30. Ibid., 3, original emphasis.

31. Ibid., 6, emphasis added.

32. See John Sekora, *Luxury* (Baltimore: Johns Hopkins University Press, 1977).

33. Tobias Smollett, *Humphry Clinker*, ed. Shaun Regan (London: Penguin, 2008), 55. All further references to this edition will be given in the text.

34. See Annika Mann, "Waste Management: Tobias Smollett and Remediation," *Eighteenth-Century Fiction* 25, no. 2 (2012–13), 359–82, esp. 380–82.

35. On sensibility and its resemblance to scientific investigation, see Ann Jessie van Sant, *Eighteenth-Century Sensibility and the Novel* (Cambridge: Cambridge University Press, 1993). Van Sant examines physiology and psychology rather than chemistry but notes the importance of the process of testing Clarissa (60–82).

36. Marcet, *Conversations on Chemistry*, 6.

37. See Leacock's discussion, in "Character, Silence, and the Novel," of the character of Ottilie in Goethe's *Elective Affinities*, whom Benjamin saw as not possessing a character because of her failure to make or articulate a decision.

38. Frances Burney, *Camilla*, ed. Edward Bloom and Lillian Bloom (Oxford: Oxford University Press, 1999), 232. All further references to this edition will be given in the text.

39. See Kristen Pond, "'Fairest Observers' and 'Restless Watchers': Contested Sites of Epistemology in Frances Burney's *Camilla*," *Studies in the Novel* 50, no. 3 (2018): 315–35.

40. On Priestley, Radcliffe, air, atmosphere, medium, and language, see Jayne Lewis, *Air's Appearance* (Chicago: University of Chicago Press, 2012).

41. Richard Kirwan, *An essay on phlogiston, and the constitution of acids. A new edition. By R. Kirwan, Esq. Member of the Academies of Stockholm Upsal, Dijon, Dublin, Philadelphia, Manchester, &c. To which are added, notes, exhibiting and defending the antiphlogistic theory; and annexed to the French edition of this work; by Messrs. de Morveau, Lavoisier, de la Place, Monge, Berthollet, and de Fourcroy: translated into English. With additional remarks and replies, by the author* (London: Johnson, 1789), 15–16.

42. Reill, *Vitalizing Nature*, 80.

43. Kirwan, *Essay on phlogiston*, v–vi.

44. Marcet, *Conversations on Chemistry*, 33.

45. Kirwan, *Essay on phlogiston*, 46.

46. Marcet, *Conversations on Chemistry*, 35.

47. Ibid.

48. Ibid., 17–19.

49. Joseph Priestley, *Experiments and observations on different kinds of air. Vol. II. By Joseph Priestley, LL.D. F.R.S.* (London: Johnson, 1775), 2:34.

50. Ibid., 35.

51. Ibid., 37.

52. Ibid., 31.

53. Kirwan, *Essay on phlogiston*, 21, emphasis added.

54. Jayne Lewis connects Radcliffe's representations to Priestley's experiments and argues, "Air carries everything in Udolpho . . . threatens to erase the distinctions among plots, characters, places, and properties. Each looks to be nothing but a temporary composition of the same elements that compose everything else" (229).

Chapter 6

1. Jane Austen, *Pride and Prejudice*, ed. James Kinsley (Oxford: Oxford University Press, 2008), 54. All further references to this edition will be given in the text.

2. Jane Austen, *Emma*, ed. James Kinsley, intro. Adela Pinch (Oxford: Oxford University Press, 2008), 186. All further references to this edition will be given in the text.

3. William Galperin, *The History of Missed Opportunities* (Stanford: Stanford University Press, 2017), 91–99. Galperin calls Miss Bates's speeches "the best 'letters' Austen ever wrote" (95). I am indebted both to this study and to his earlier book, William Galperin, *The Historical Austen* (Philadelphia: University of Pennsylvania Press, 2005).

4. Robert Brown, "A Brief Account of Microscopical Observations Made in the Months of June, July, and August 1827, on the Particles Contained in the Pollen of Plants; and on the General Existence of Active Molecules in Organic and Inorganic Bodies," *Philosophical Magazine and Annals of Philosophy* (London, Richard Taylor), 4, no. 21: 161–73, 166. Brown notes that "the dust or soot deposited on all bodies in such quantity, especially in London, is entirely composed of these molecules" (166), but his focus remains on the process of fertilization in ways that resonate with the plot of *Emma*.

5. Jean Perrin, *Brownian Movement and Molecular Reality*, trans. F. Soddy (London: Taylor and Francis, 1910), 1–2, original emphasis.

6. Ibid., 3. In this there is a striking resemblance to the argument about the everyday in Galperin, *History of Missed Opportunities*.

7. Perrin, *Brownian Movement and Molecular Reality*, 1–2, 7, 6, original emphases.

8. Peter Knox-Shaw, *Jane Austen and the Enlightenment* (Cambridge: Cambridge University Press, 2004), 24.

9. As in D. A. Miller, *Narrative and Its Discontents* (Princeton: Princeton University Press, 1981). See Brian Richardson, *A Poetics of Plot for the Twenty-First Century* (Columbus: Ohio State University Press, 2019), for a full and recent discussion of plot.

10. I build on but also depart somewhat from Tita Chico's argument in "Minute Particulars: Microscopy and Eighteenth-Century Narrative," *Mosaic* 39, no. 2 (2006): 143–61.

11. Perrin, *Brownian Movement and Molecular Reality*, 6. He notes that the study of thermodynamics earlier in the century had resulted in the widespread acceptance of the molecular hypothesis, but that Brown's discovery not only confirmed (or anticipated) this empirically, but also suggested new things about the movements of molecules.

12. Walter Scott, "Emma: A Novel," *Quarterly Review* 14 (October 1815): 188–201, 197. On the comparison of novels with Flemish painting, see Ruth Yeazell, *Art of the Everyday* (Princeton: Princeton University Press, 2008).

13. See Peter Brooks, *Reading for the Plot* (New York: Random House, 1984). More recently, see David Kurnick, "Jane Austen, Secret Celebrity, and Mass Eroticism," *New Literary History* 52, no. 1 (2021): 53–75. I am also indebted to an earlier version of Kurnick's essay.

14. Claudia Johnson argues compellingly that "taking note of [objects] unduly . . . marks us as outsiders to that world . . . whenever objects are made to stand out with any sort of specificity in Austen's novels, something is wrong" (*Jane Austen's Cults and Cultures* [Chicago: University of Chicago Press, 2012], 161–63). Johnson concludes that "it is relations rather than things . . . that compel" Austen (171).

15. Austen, *Persuasion*, ed. James Kinsley, intro. Deidre Lynch (Oxford: Oxford University Press, 2004), 52. All further references to this edition will be given in the text.

16. On the impersonality of style, see D. A. Miller, *Jane Austen, or the Secret of Style* (Princeton: Princeton University Press, 2003); Frances Ferguson, "Now It's Personal: D. A. Miller and Too-Close Reading," *Critical Inquiry* 41, no. 3 (2015): 521–40.

17. Deidre Lynch, *Loving Literature: A Cultural History* (Chicago: University of Chicago Press, 2014).

18. Willem Jacob s'Gravesande, *Mathematical Elements of Natural Philosophy Confirmed by Experiments, or an Introduction to Newton's Philosophy*, trans. J. T. Desaguliers, 2nd ed. (London: Taylor, 1721). "Of Body in General," s'Gravesande notes, "We acquire an Idea of Solidity by the Touch: We feel that some Bodies resist us . . . which Idea of Solidity we transfer to those more subtle Bodies, which, by reason of the Smallness of their Parts, escape our Senses; and we find by experience, that even those resist other Bodies" (6).

19. Austen, *Emma*, 320; Austen, *Persuasion*, 29. All further references to both editions will be given in the text.

20. For example, see Jerome Bruner's essay, "The Narrative Construction of Reality," *Critical Inquiry* 18, no. 1 (1991): 1–21. See also Gerald Prince, *A Grammar of Stories* (The Hague: Mouton, 1973).

21. Jane Austen, *Mansfield Park*, ed. Kathryn Sutherland (London: Penguin, 1996), 346. All further references to this edition will be given in the text.

22. Deidre Lynch, "At Home with Jane Austen," in *Cultural Institutions of the Novel*, ed. Deidre Lynch and William Warner (Durham, NC: Duke University Press, 1996), 159–92, 163. Lynch, *Loving Literature*, 187.

23. Knox-Shaw, *Jane Austen and the Enlightenment*, 20.

24. Jan Golinski, *Science as Public Culture* (Cambridge: Cambridge University Press, 1992), 259.

25. Quoted in Knox-Shaw, *Jane Austen and the Enlightenment*, 23. See Galperin, *The Historical Austen*.

26. Knox-Shaw, *Jane Austen and the Enlightenment*, 23.

27. Jane Austen, *Northanger Abbey*, ed. Marilyn Butler (London: Penguin, 1995), 37.

28. Daniel Defoe, *The Storm*, ed. Richard Hamblyn (New York: Penguin, 2005), 14.

29. Steven Shapin, *The Scientific Revolution* (Chicago: University of Chicago Press, 1996), 46, 53.

30. Alex Woloch, *The One vs. the Many* (Princeton: Princeton University Press, 2003), 12.

31. s'Gravesande, *Mathematical Elements*, 2.

32. See Frances Ferguson, "Jane Austen, Emma, and the Impact of Form," *MLQ* 61, no. 1 (2000): 157–80; and Miller, *Jane Austen, or the Secret of Style*.

33. Ian Hacking, *The Taming of Chance* (Cambridge: Cambridge University Press, 1990), 112.

34. My argument owes much also to the growing body of work on "embodied consciousness" in Austen; see Alan Richardson, *British Romanticism and the Science of Mind* (Cambridge: Cambridge University Press, 2001), and Alan Richardson, *The Neural Sublime* (Baltimore: Johns Hopkins University Press, 2010); Kay Young, "Feeling Embodied: Consciousness, Persuasion, and Jane Austen," *Narrative* 11, no. 1 (2003): 78–92, and Kay Young, *Imagining Minds* (Columbus: Ohio State University Press, 2010); Antonina Harbus, "Reading Embodied Consciousness in Emma," in *SEL* 51, no. 4 (2011): 765–82. Their focus is defined somewhat differently, however, in terms of psychological and philosophical problems.

35. Benjamin Martin, *A plain and familiar introduction to the Newtonian philosophy* (London: Owen, 1754), 157; Shapin, *Scientific Revolution*, 46.

36. Knox-Shaw, *Jane Austen and the Enlightenment*, 13.

37. Quoted in Knox-Shaw, *Jane Austen and the Enlightenment*, 12.

38. Peter Hanns Reill, *Vitalizing Nature in the Enlightenment* (Berkeley: University of California Press, 2005), 80.

39. s'Gravesande, *Mathematical Elements*, 21.

40. Martin, *Plain and familiar introduction*, 157.

41. Shapin, *Scientific Revolution*, 46.

42. John Desaguliers, *A Course of Experimental Philosophy* (London: Longman, 1734), 1:6.

43. s'Gravesande, *Mathematical Elements*, 49–50. This passivity of matter is a key feature of the mechanical philosophy, and Austen's plots often foreground a kind of baseline inertia as a rule of material being. For a different interpretation of passivity and a theory of "recessive action," see Anne-Lise François, *Open Secrets* (Stanford: Stanford University Press, 2008). I agree with François that *Mansfield Park* seems to prefer reticence and resistance, but preference and explanation have different functions. Feminist criticism of Austen is both historically deep and broad; see the essays in *Janeites*, ed. Deidre Lynch (Princeton: Princeton University Press, 2000). See also Claudia Johnson, *Jane Austen: Women, Politics, and the Novel* (Chicago: University of Chicago Press, 1990); Clara Tuite, *Romantic Austen* (Cambridge: Cambridge University Press, 2002); and, particularly relevant to this chapter's argument, Jillian Heydt-Stevenson, *Austen's Unbecoming Conjunctions* (New York: Palgrave, 2005).

44. Desaguliers, *Course of Experimental Philosophy*, 42.

45. Martin, *Plain and familiar introduction*, 157.

46. On plot as delay, see Brooks, *Reading for the Plot*. The difference is that Brooks grounds this delay in a Freudian model of the psyche rather than a model of the physical world. On friction, see Anna Lowenhaupt Tsing, *Friction* (Princeton: Princeton University Press, 2005).

47. See Young, "Feeling Embodied," for a reading of this moment.

48. Georges Louis Leclerc, Comte de Buffon. *The natural history of animals, vegetables, and minerals; with the theory of the earth in general. Translated from the French of Count de Buffon. Intendant of the Royal Gardens in France; Member of the French Academy, of the Academy of Sciences, and of the Royal Societies of London Berlin, &c. By W. Kenrick, L.L.D. and J. Murdoch*, vol. 1 (London: Bell, [1775]–76), 397–98.

49. For the classic reading of the novel as a critique of Emma's subjectivity, see Marilyn Butler, *Jane Austen and the War of Ideas* (Oxford: Clarendon, 1975).

50. See E. M. Forster, *Aspects of the Novel* (San Diego: Harcourt Brace, 1985).

51. See William Galperin, "Describing What Never Happened: Jane Austen and the History of Missed Opportunities," *ELH* 73, no. 2 (2006): 355–82.

Chapter 7

1. Maria Edgeworth, *Belinda*, ed. Kathryn Kirkpatrick (Oxford: Oxford University Press, 1994), 32. All further references to this edition will be given in the text.

2. To cite an older work and a recent one, see Peter Brooks, *Reading for the Plot* (New York: Vintage, 1984); Ian Duncan, *Human Forms* (Princeton: Princeton University Press, 2019). Duncan's subtle argument emphasizes how, in partnership with the emerging sciences of the human, novels explored human nature as "variable, fluid, fleeting" rather than fixed (3).

3. Michel Foucault, *The Order of Things* (New York: Random House, 1973), 46–50.

4. Foucault, *The Order of Things*, 50.

5. I use the term "transitional object" very loosely. See Vanessa Smith, "Toy Stories," *Novel: A Forum on Fiction* 50, no. 1 (2017): 35–55 on the original context of use.

6. Lorraine Daston and Peter Galison, *Objectivity* (New York: Zone Books, 2007).

7. I understand plot to consist of both the sequence of events and the order of their disclosure; that is, I see it operating on both levels of what narrative theorists call story and discourse. The narrator I take to be the agent in charge only of the latter. On the omniscient narrator, see Audrey Jaffe, *Vanishing Points* (Berkeley: University of California Press, 1991).

8. O. B. Hardison and E. H. Behler, "Topos," in *The Princeton Encyclopedia of Poetry and Poetics*, ed. Roland Green, Stephen Cushman, and Clare Cavanagh, 4th ed. (Princeton: Princeton University Press, 2012).

9. The argument in this chapter is indebted to many excellent studies of attention in the eighteenth and nineteenth centuries, including Caroline Levine, *The Serious Pleasures of Suspense* (Charlottesville: University of Virginia Press, 2003); Mary Favret, *War at a Distance* (Princeton: Princeton University Press, 2010); Lily Gurton-Wachter, *Watchwords* (Stanford: Stanford University Press, 2016); Natalie Phillips, *Distraction* (Baltimore: Johns Hopkins University Press, 2016); see also Luke Terlaak Poot, "On Cliffhangers," *Narrative* 24, no. 1 (2016): 50–67.

10. *Aristotle's "Poetics,"* trans. James Hutton (New York: Norton, 1982), 56.

11. Although their cultural status diminishes, as Terence Cave has argued in his study, *Recognitions* (Oxford: Clarendon, 1990).

12. James Chandler, *England in 1819* (Chicago: University of Chicago Press, 1998), 101. Nicholas Paige disputes the idea that the change was an epistemic one, but also asserts a change "around the turn of the nineteenth century" in "people's literary behavior"; he locates the change in a shift to third-person narration and the abandonment of the pseudofactual claim (*Before Fiction* [Philadelphia: University of Pennsylvania Press, 2011], 26).

13. Foucault, *The Order of Things*, 76.

14. Ibid., 120.

15. Ibid., 133–34.

16. Ibid., 268.

17. Ibid.

18. The painting provides a visual analogue for Foucault's argument and the subject of his first chapter in *The Order of Things* (3–16).

19. Foucault, *The Order of Things*, 265, 259.

20. Ibid., 220.

21. For a compelling (and complementary) account of "interiority as an effect of public and social discourses" and the development of the market, see Deidre Lynch, *The Economy of Character* (Chicago: University of Chicago Press, 1998), 168.

22. Foucault, *The Order of Things*, xxii.

23. Georges Cuvier, "Preliminary Discourse," in Martin Rudwick, *Georges Cuvier, Fossil Bones, and Geological Catastrophes* (Chicago: University of Chicago Press, 1997), 217.

24. Foucault, *The Order of Things*, 239.

25. Daston and Galison, *Objectivity*, 59.

26. Ibid., 53.

27. Ibid., 63.

28. Ibid., 59.

29. Ibid., 37. They note that Bacon's famous warning "had nothing to do with the suppression of the subjective self, but rather addressed the balance between opposing tendencies to excess" (32).

30. Daston and Galison, *Objectivity*, 121.

31. Ibid., 184–85.

32. See Michael Fried, *Absorption and Theatricality* (Berkeley: University of California Press, 1980).

33. Ibid., 201. Kant also provides a critical hinge for Foucault's argument about the shift from the classical to the modern episteme.

34. Schelling names *Don Quixote* as one of the only two novels written so far, alongside Goethe's *Wilhelm Meister*, in Friedrich Wilhelm Joseph von Schelling, *Philosophy of Art*, trans. Douglas W. Stott (Minneapolis: University of Minnesota Press, 1985), 234–35. On the context of Cervantes's work, see Rachel Schmidt, *Forms of Modernity* (Toronto: University of Toronto Press, 2011); and David Quint, *Cervantes's Novel of Modern Times* (Princeton: Princeton University Press, 2003). The most notable modern treatment of *Don Quixote* as an archetypal novel can be found in Georg Lukács's *Theory of the Novel*, trans. Anna Bostock (Cambridge, MA: MIT Press, 1983).

35. In other contexts besides the history of science, scholars have studied how the idea of subjectivity became both prominent and problematic around the end of the eighteenth century, as borne out in novels in particular. In "The Novel and Prejudice," Sarah Winter has argued that, starting around 1800, there emerges a subgenre of novels that "overtly engage prejudice as

both an automatic mode of thought and an expression and experience of bias" (90) that can be overcome by rational reflection and conscience. They include several discussed in this chapter. See Sarah Winter, "The Novel and Prejudice," *Comparative Literature Studies* 46, no. 1 (2009): 76–102. The emphasis on the perils of subjectivity in novels of this period has also been well studied in connection to the French Revolution and the British response; see, for example, Marilyn Butler, *Romantics, Rebels, and Reactionaries* (Oxford: Oxford University Press, 1981); Claudia Johnson, *Equivocal Beings* (Chicago: University of Chicago Press, 1995). I make a somewhat different argument in Yoon Sun Lee, *Nationalism and Irony* (New York: Oxford University Press, 2004). While developments in natural philosophy were not completely insulated from this historical context, I want to suggest a more complex mediation. On the struggle to attain objectivity in the Victorian context along the lines described by Daston and Galison, see George Levine, *Dying to Know* (Chicago: University of Chicago Press, 2002).

36. Henry Fielding, *Don Quixote in England: A Comedy* (London: J. Watts, 1734), 63. Its final song states it explicitly: "All Mankind are mad, 'tis plain; Some for Places, Some Embraces; Some are mad to keep up Gain, And others mad to spend it" (63). On the reception history, see Ronald Paulson, *Don Quixote in England* (Baltimore: Johns Hopkins University Press, 1998).

37. Daston and Galison, *Objectivity*, 203, emphasis added.

38. A different approach to the question of representing private or interior spaces can be found in Michael McKeon, *The Secret History of Domesticity* (Baltimore: Johns Hopkins University Press, 2005).

39. For fuller examinations of quixotism, see Paulson, *Don Quixote in England*; Wendy Motooka, *The Age of Reasons* (London: Routledge, 1998); Aaron Hanlon, *A World of Disorderly Notions* (Charlottesville: University of Virginia Press, 2019).

40. Henry Fielding, *Joseph Andrews*, ed. Homer Goldberg (New York: Norton, 1987), 18.

41. Ibid.

42. Sarah Fielding and Jane Collier, *The Cry*, ed. Carolyn Woodward (Lexington: University Press of Kentucky, 2018), 290.

43. Foucault, *The Order of Things*, 268. Even the inner struggles of Lady Booby, for example, are directly represented as a theatrical dialogue between the various passions within her mind; there is no need to have her feelings drawn out through events: "Love became [Joseph's] Advocate, and whispered many things in his favour . . . on the other side, Pride and Revenge spoke as loudly against him: and thus the poor Lady was tortured with Perplexity; opposite Passions distracting and tearing her Mind different ways" (36).

44. Amanda Gilroy, introduction to *The Female Quixote*, by Charlotte Lennox, ed. Amanda Gilroy and Wil Verhoeven (London: Penguin, 2006), xii. See the reading of this novel in Catherine Gallagher, *Nobody's Story* (Berkeley: University of California Press, 1994).

45. Cervantes's novel also presents Don Quixote's cure as an arbitrary authorial decision, made more to preempt unauthorized continuations of his adventures than for any internal reason.

46. My thinking has benefited from the work of Frances Ferguson on free indirect discourse, its implications and its context; see in particular Frances Ferguson, "Jane Austen, Emma, and the Impact of Form," *MLQ* 61, no. 1 (2000): 157–80; and Frances Ferguson, "Not Kant, but Bentham," *Critical Inquiry* 45, no. 3 (2019): 577–600.

47. On the physiology of novel-reading, see Nicholas Dames, *The Physiology of the Novel* (New York: Oxford University Press, 2007). On the belief in the therapeutic value of book-reading, see Leah Price, *What We Talk About When We Talk About Books* (New York: Basic Books, 2019).

48. See Richard Sha, *Imagination and Science in Romanticism* (Baltimore: Johns Hopkins University Press, 2018).

49. Jan Golinski, *The Experimental Self* (Chicago: University of Chicago Press, 2016), 101.

50. Humphry Davy, "A Discourse Introductory to a Course of Lectures on Chemistry, Delivered in the Theatre of the Royal Institution on the 21st of Jan, 1802," in *The Collected Works of Sir Humphry Davy*, ed. John Davy (London: Smith, Elder, 1839), 2:307–26, 315.

51. Humphry Davy, *Researches, chemical and philosophical; chiefly concerning nitrous oxide, or dephlogisticated nitrous air, and its respiration* (London: printed for J. Johnson, by Biggs and Cottle, Bristol, 1800), 454.

52. Ibid., 458–59.

53. Ibid., 463.

54. Ibid., 460, 479.

55. Ibid., 488–89. See M. Jay, "The Atmosphere of Heaven: The 1799 Nitrous Oxide Researches Reconsidered," *Notes and Records of the Royal Society of London* 63, no. 3 (2009): 297–309, as well as Golinski, *Experimental Self*.

56. Davy, *Researches*, 489, 493.

57. Quoted in Golinski, *Experimental Self*, 32.

58. Though not all visitors experienced the same effects from breathing the gas, most of them, including Richard Lovell Edgeworth, did. Maria Edgeworth was skeptical, inclined to attribute the effects of the gas only to "faith, great faith." Quoted in Jay, "Atmosphere of Heaven," 300; Golinski, *Experimental Self*, 37.

59. "Davy continued to cultivate an embodied mode of investigation that relied on self-experimentation and public self-display . . . Davy's reputation as a genius remained associated with a willingness to subject his body to strenuous trials in the course of his experimental inquiries" (Golinski, *Experimental Self*, 45).

60. Davy, "A Discourse Introductory," 307–26, 311–12.

61. Ibid., 311, emphasis added.

62. Jane Austen, *Persuasion*, ed. James Kinsley (Oxford: Oxford University Press, 2004), 53–54.

63. Clara Reeve, *The Progress of Romance* (Colchester: W. Keymer, 1785), 1:111.

64. Walter Scott, "Emma: A Novel," *Quarterly Review* 14 (October 1815): 188–201, 193.

65. Davy, "A Discourse Introductory," 326.

66. Ibid., 326, emphasis added.

67. Ibid., 319, 326.

68. Dames, *Physiology of the Novel*, 47, 57.

69. Jane Austen, *Northanger Abbey*, ed. James Kinsley (Oxford: Oxford University Press, 2003), 77–78. On suspense and the Gothic, see Phillips, *Distraction*, 162–65.

70. On the rich context of Romantic science, see, for example, Sharon Ruston, *Creating Romanticism* (New York: Palgrave, 2013); Jon Klancher, *Transfiguring the Arts and Sciences* (Cambridge: Cambridge University Press, 2013); Robert Mitchell, *Experimental Life* (Baltimore: Johns Hopkins University Press, 2013).

71. See Norma Clarke, "The Cursed Barbauld Crew," in *Opening the Nursery Door*, ed. Mary Hilton, Morag Styles, and Victor Watson (New York: Routledge, 1997), 91–103.

72. Frances Ferguson, "The Novel Comes of Age," *differences* 28, no. 1 (2017): 37–63, 37.

73. Ibid., 57.

74. From *Waverley* (1814), to *The Heart of Midlothian* (1818), to *Redgauntlet* (1824).

75. Maria Edgeworth, *Harry and Lucy Concluded, Being the Last Part of "Early Lessons" by Maria Edgeworth*, 3rd ed., revised and corrected, 3 vols. (London: Baldwin and Cradock, 1837), 1:ix. Further references to this edition are given in the text.

76. Maria Edgeworth, *Early Lessons*, 3 vols. (Paris: Baudry's European Library, 1847), 3–76. Further references to this edition are given in the text.

77. See Aileen Douglas, "Time and the Child," in *Children's Literature Collections*, ed. K. O'Sullivan and P. Whyte (New York: Palgrave Macmillan, 2017), 91–105, https://doi.org/10.1057/978-1-137-59757-1_6.

78. Edgeworth, *Harry and Lucy Concluded*, 3rd ed., 3 vols. (1837), 1:213–14.

79. Maria Edgeworth, *Harry and Lucy Concluded*, 4 vols. (London: Baldwin and Cradock, 1825), 2:108. Further references to this edition will be given in the text.

80. Maria Edgeworth, *Harry and Lucy*, 3 vols. (London: Baldwin and Cradock, 1840), 3:250–51. Further references to this edition are given in the text. On these two types of distraction, see Phillips, *Distraction*.

81. Maria Edgeworth, *Early Lessons*, 12th ed., 4 vols. (London: Baldwin and Cradock, 1833), 2:294.

82. The intertextuality is remarkable; Barbauld's stories feature a character named Lucy, for example.

83. Anna Laetitia Barbauld and John Aikin, *Evenings at Home*, vol. 4 (London: Johnson, 1794), 108–9.

84. Reading this account in the context of slavery, as we must, produces a much more chilling effect; see Keith Sandiford, *The Cultural Politics of Sugar* (Cambridge: Cambridge University Press, 2000); Arthur Stinchcombe, *Sugar Island Slavery in the Age of Enlightenment* (Princeton: Princeton University Press, 1995); Kay Dian Kriz, *Slavery, Sugar, and the Culture of Refinement* (New Haven: Yale University Press, 2008).

85. Daston and Galison, *Objectivity*, 139.

86. See Terlaak Poot, "On Cliffhangers."

87. Edgeworth, *Early Lessons*, 12th ed., 4 vols., 57.

88. See James Chandler, "Edgeworth and Realism," unpublished/forthcoming ms.; as well as Nicole Wright, "Opening the Phosporic 'Envelope': Scientific Appraisal, Domestic Spectacle, and (un)'Reasonable Creatures' in Edgeworth's *Belinda*," *Eighteenth-Century Fiction* 24, no. 3 (2012): 509–36.

89. See Cave, *Recognitions*.

90. See Wendy Moore, *How to Create the Perfect Wife* (New York: Basic Books, 2013).

91. Edgeworth adheres here to the French usage, conflating experience and experiment. See John Heilbron, "Natural Philosophy," in *Wrestling with Nature*, ed. Peter Harrison, Ronald L. Numbers, and Michael H. Shank (Chicago: University of Chicago Press, 2011), 173–99.

92. Jonathan Kramnick, *Actions and Objects from Hobbes to Richardson* (Stanford: Stanford University Press, 2010), 11.

93. In *Belinda*, it will reveal, with the help of Clarence (and a gaming table), the truth of Vincent's character; see Yoon Sun Lee, "Bad Plots and Objectivity in Maria Edgeworth," *Representations* 139 (2017): 34–59, 52.

94. See Claudia Brodsky, "Austen's *Pride and Prejudice* and Hegel's 'Truth in Art': Concept, Reference, and History," *ELH* 59, no. 3 (1992): 597–623.

95. The printed advertisement that Falkland puts out that names and describes Williams, however, seems to pursue him into every corner of England; it functions in ways very similar to the objects in Edgeworth's tales and novels.

96. William Godwin, *Caleb Williams*, ed. Maurice Hindle (London: Penguin, 1988), 349–51.

97. William Godwin, *Enquiry Concerning Political Justice*, 3rd ed., 2 vols. (London: Robinson, 1798), 2:420–21. Godwin argues somewhat equivocally, "There is no such thing as action. Man is in no case . . . the beginner of any event or series of events . . . but only the vehicle through which certain antecedents operate. . . . Action however, in its more simple and obvious sense, is sufficiently real, and exists equally both in mind and in matter" (2:385).

98. Ibid., 401.

99. "a-poros, a path that is blocked, is not the same as a non-path, a path never opened," Alexander Nagel and Lorenzo Pericolo, "Unresolved Images: An Introduction to Aporia as an Analytical Category in the Interpretation of Early Modern Art," in *Subject as Aporia in Early Modern Art*, ed. Alexander Nagel and Lorenzo Pericolo (New York: Routledge, 2010), 1–16, 9.

100. Maria Edgeworth, *Harrington*, ed. Susan Manly (Ontario, Canada: Broadview, 2004), 71, 91. All further references to this edition are given in the text. On the role of the nurse in this period and in *Harrington* in particular, see Katie Trumpener, *Bardic Nationalism* (Princeton: Princeton University Press, 1997), 193–241.

101. One is Harrington's father, who swears "by Jupiter Ammon" that his son would never "marry a Jewess" (193). It is the novel's peculiar conceit that this oath, once sworn, "seemed to bind even him who uttered [it] beyond his own power of revocation" (87). It becomes an objective fact or reality and can only be attenuated by a series of plot events and, finally, circumvented by the revelation, disappointing to modern readers, that Berenice's mother was Christian.

102. Edgeworth alludes to the work of Kenelm Digby, a seventeenth-century natural philosopher and alchemist who believed, notably, in sympathetic cures. See Seth Lobis, "Sir Kenelm Digby and the Power of Sympathy," *Huntington Library Quarterly* 74, no. 2 (2011): 243–60.

103. Jane Austen, *Pride and Prejudice*, ed. Tony Tanner (London: Penguin, 1985), 236. All further references to this edition are given in the text.

104. Jane Austen, *Emma*, ed. James Kinsley (Oxford: Oxford University Press, 2008), 263. All further references to this edition are given in the text.

105. On the use of free indirect discourse in this novel, see Ferguson, "Jane Austen, Emma, and the Impact of Form"; Daston and Galison, *Objectivity*, 160.

106. The discovery is a lock of hair that slips out of a letter that Clarence Hervey holds, in an allusion, presumably, to Belinda's namesake in Pope's poem. Belinda had already been engaged in a "self-examination," recollecting her various reactions to things Hervey had said or done. "She was sensible that the sight of a lock of hair . . . in the hands of any man but Clarence Hervey, could not possibly have excited any emotion in her mind" (139).

107. Daston and Galison, *Objectivity*, 185.

Chapter 8

1. Walter Scott, "General Preface to the Waverley Novels," in *Waverley*, ed. Claire Lamont (Oxford: Oxford University Press, 1986), 350. Scott remarks that Waverley's education was based on his own; in his case, though, it was a circulating library in Edinburgh, from which he borrowed "specimens of every kind [of works of fiction], from the romances of chivalry, and

the ponderous folios of Cyrus and Cassandra, down to the most approved works of later times" (350). All further references to this edition will be given in the text.

2. One of Scott's favorite varieties of quixote is the antiquarian, as in the character of Jonathan Oldbuck. On the ideological functions of antiquarianism, see Yoon Sun Lee, "A Divided Inheritance: Scott's Antiquarian Novel," *ELH* 64, no. 2 (1997): 537–67.

3. On the influence of Scott and the circumstances surrounding his fame, see Ian Duncan, *Scott's Shadow* (Princeton: Princeton University Press, 2007).

4. See Ian Duncan, "Walter Scott and the Historical Novel," in *English and British Fiction 1750–1820*, ed. Peter Garside and Karen O'Brien, vol. 2 of *The Oxford History of the Novel in English* (Oxford: Oxford University Press, 2015), 312–34. Duncan observes, "The novel's mixed stylistic textures . . . [render] a historical and regional diversity of human communities, which it curates as though in a print museum" (320).

5. Although I will not pursue this line, on the important discourses and debates surrounding natural historical extinction, particularly in Buffon and Cuvier, see Martin J. S. Rudwick, *Georges Cuvier, Fossil Bones and Geological Catastrophes: New Translations and Interpretations of the Primary Texts* (Chicago: University of Chicago Press, 1997); Phillip Sloan, "Kant on the History of Nature," *Studies in History and Philosophy of Biological and Biomedical Sciences* 37 (2006): 627–48; Joanna Stalnaker, "Buffon on Death and Fossils," *Representations* 115, no. 1 (2011): 20–41. On the development and extinction of species, see Gillian Beer, *Darwin's Plots* (Cambridge: Cambridge University Press, 2000); on the development as key to the conceptualization of both the human and the novel, see Ian Duncan, *Human Forms* (Princeton: Princeton University Press, 2020).

6. G. W. F. Hegel, *Aesthetics*, trans. T. M. Knox, 2 vols. (Oxford: Clarendon, 1975), 2:1090. On the relation between history and ethnography in Romantic historicism, see James Chandler, *England in 1819* (Chicago: University of Chicago Press, 1998), 105–9.

7. Here I would respectfully disagree with those who see Scott's novels as evoking an "end of history," as in Jerome Christensen's *Romanticism at the End of History* (Baltimore: Johns Hopkins University Press, 2000). The message in the Waverley novels seems to be that the things you thought were long since settled are in fact never settled but still alive with potential.

8. The pardon was made necessary because of a recent legal change that was in turn necessitated by a particular political situation.

9. Georg Lukács, *The Historical Novel*, trans. Hannah Mitchell and Stanley Mitchell (Lincoln: University of Nebraska Press, 1983), 52. In this he sees proof of the historical novel's epic character. Such characters only discover their heroism because of the objective circumstances in which they find themselves: in real life, "Endless numbers of people live out their lives quietly, without this upsurge, because no opportunity has come their way. . . . Revolutions are thus the great periods of mankind because in and through them such rapid upward movements in human capacities become widespread" (53).

10. Walter Scott, *Old Mortality*, ed. Angus Calder (London: Penguin, 1985), 186. All further references to this edition will be given in the text.

11. Walter Scott, "Emma: A Novel," *Quarterly Review* 14 (October 1815): 188–201, 192.

12. Ibid., 192–93.

13. Willem Jacob s'Gravesande, *Mathematical Elements of Natural Philosophy Confirmed by Experiments, or an Introduction to Newton's Philosophy*, trans. J. T. Desaguliers (London: Longman, 1721), 2.

14. Benjamin Martin, *A plain and familiar introduction to the Newtonian philosophy* (London: Owen, 1754), 138–39.

15. On vection, see the fascinating study by Alain Berthoz, *The Brain's Sense of Movement*, trans. Giselle Weiss (Cambridge, MA: Harvard University Press, 2000).

16. In Smollett's *Humphry Clinker*, movement takes on both sociological as well as physiological significance. The journey chronicled in that novel reveals the fast-urbanizing social world, uncovers dimensions of the fictive world (new or hidden relationships between characters, for example), and affects characters' inner states of being in highly divergent and memorable ways. The modern urban world is vividly represented as a phenomenon of mobility. But for the conservative Smollett, mobility as a social feature falls within the larger category of disorder; it is a concept that remains in opposition to an unchanging ideal of order. Smollett allows for the physical benefits of travel and motion, but movement still remains instrumental, a way of arriving at a condition of health (or its opposite). Most tellingly, for Smollett it cannot be a matter of doubt whether one is in motion or at rest.

17. On the role of gender, see Ina Ferris, *The Achievement of Literary Authority: Gender, History, and the Waverley Novels* (Ithaca: Cornell University Press, 1991).

18. See Alan Bewell, *Natures in Translation* (Baltimore: Johns Hopkins University Press, 2017), as well as his earlier article "Erasmus Darwin's Cosmopolitan Nature," *ELH* 76, no. 1 (2009): 19–48. Erasmus Darwin was a central member of the network that included Richard Lovell Edgeworth, whose daughter, the author Maria Edgeworth, Scott knew and corresponded with; Scott was also linked with Darwin through his acquaintance with Anna Seward and Wordsworth. In a letter to Richard Lovell Edgeworth, Scott describes Darwin as "ranking among the British poets of the highest class," despite having apparently plagiarized four lines from Anna Seward in *The Botanic Garden* (A. H. Ashe, "Two Letters of Sir Walter Scott, 1812," *Notes and Queries* 40, no. 4 [1993], 458–61, 459). See also John Brewer, *A Sentimental Murder* (New York: Farrar, Strauss and Giroux, 2004), 186–90; Devin Griffiths, *The Age of Analogy* (Baltimore: Johns Hopkins University Press, 2016); Noel Jackson, *Science and Sensation in Romantic Poetry* (Cambridge: Cambridge University Press, 2008); Alan Richardson, *British Romanticism and the Science of the Mind* (Cambridge: Cambridge University Press, 2001).

19. Scott's maternal grandfather, John Rutherford, was one of four Scottish students trained in Leiden by Hermann Boerhaave, one of the founders of neuroscience. Rutherford taught at the medical school alongside Robert Whytt, who discovered the spinal cord and the optic nerve, and William Cullen, the author of an influential nosology, among other publications. See C. U. M. Smith, "Brain and Mind in the Long Eighteenth Century," in *Brain, Mind, and Medicine*, ed. Harry Whitaker and C. U. M. Smith (Cham: Springer, 2007), 15–28; Nima Bassiri, "The Brain and the Unconscious Soul in Eighteenth-Century Nervous Physiology: Robert Whytt's Sensorium Commune," *Journal of the History of Ideas* 74, no. 3 (2013): 425–48; Julius Rocca, "William Cullen (1710–1790) and Robert Whytt (1714–1766) on the Nervous System," in *Brain, Mind, and Medicine*, ed. Whitaker and Smith, 85–98; Gavin Budge, *Romanticism, Medicine and the Natural Supernatural* (New York: Palgrave, 2013).

20. Erasmus Darwin, *Zoonomia; or, the laws of organic life.* [. . .]. *By Erasmus Darwin, M.D. F.R.S. Author of The Botanic Garden*, 2nd ed. (London: Johnson, 1796), 1:1. All further references to this edition will be given in the text.

21. Jacques Rancière, "The Thread of the Novel," *Novel: A Forum on Fiction* 47, no. 2 (2014): 196–209, 203–4.

22. Scott, "Emma," 193.

23. On the importance of such feats to the novel, see Margaret Cohen, *The Novel and the Sea* (Princeton: Princeton University Press, 2010).

24. On spectral appearances, see Ina Ferris, "'Before Our Eyes': Romantic Historical Fiction and the Apparitions of Reading," *Representations* 121 (2013): 60–84.

25. Compare this passage from the beginning of *Waverley*, when the protagonist seeks out the dark, empty library to enjoy "that internal sorcery by which past or imaginary events are presented in action, as it were, to the eye of the muser . . . listening to the decaying echo of the hoofs of the king's horse, and when that had died away. . . . A distant sound is heard like the rushing of a swoln stream; it comes nearer, and Edward can plainly distinguish the galloping of horses . . . straggling pistol-shots between, rolling forwards to the hall" (17). The acoustic effects are remarkable in this scene of what we might call self-induced vertigo.

26. Martin, *Plain and familiar introduction*, 139, 142.

27. It is not that Darwin offers a Lamarckian argument about the heritability of such catenations, though Scott plays with such an idea. On British interest in Lamarck and evolutionary ideas in late Scott and Victor Hugo, see Duncan, *Human Forms*, 86–122.

28. Alexander Welsh, *The Hero of the Waverley Novels* (Princeton: Princeton University Press, 2014).

29. Scott nearly always includes self-interest and prudential calculation among these feelings; it works like a force of gravity that turns the smaller wheels. It's presented more as an instinct or habit than as an exercise of reason.

30. Walter Scott, *The Heart of Midlothian*, intro. David Daiches (New York: Rinehart, 1957), 32. All further references to this edition will be given in the text.

31. In his *Antinomies of Realism* (London: Verso, 2013), Fredric Jameson cites a scene that occurs immediately after this one to show how Scott's novel represents collectivity: "The historical novel as a genre cannot exist without this dimension of collectivity, which marks the drama of the incorporation of the individual characters into a greater totality, and can alone certify the presence of History as such" (267).

32. On crowd-forms and the agency of the crowd, see John Plotz, *The Crowd: British Literature and Public Politics* (Berkeley: University of California Press, 2000).

33. Walter Scott, *The Bride of Lammermoor*, ed. Fiona Robertson (Oxford: Oxford University Press, 1991), 171. All further references to this edition will be given in the text.

34. On the counterfactual novel, see Catherine Gallagher, *Telling It Like It Wasn't* (Chicago: University of Chicago Press, 2018).

35. See Christensen, *Romanticism at the End of History*.

36. Walter Scott, *Redgauntlet*, ed. Kathryn Sutherland (Oxford University Press, 1985), 33. All further references to this edition will be given in the text.

37. Lukács, *The Historical Novel*, 144.

38. Erich Auerbach, *Mimesis* (Princeton: Princeton University Press, 1974), 443.

39. Ibid. On historicism, see Chandler, *England in 1819*; on the problem of sensing history, see Kevis Goodman, "'Uncertain Disease': Nostalgia, Pathologies of Motion, Practices of Reading," *Studies in Romanticism* 49 (2010): 197–227; Mary Favret, *War at a Distance: Romanticism and the Making of Modern Wartime* (Princeton: Princeton University Press, 2017); Lily Gurton-Wachter, *Watchwords* (Stanford: Stanford University Press, 2016).

40. Auerbach, *Mimesis*, 447.

41. Ibid. It is not until Marx, he affirms (the only appearance that Marx makes in Auerbach's book), that we see "the presently visible germs of the concrete future" (445).

Epilogue

1. See Katie Trumpener, *Bardic Nationalism* (Princeton: Princeton University Press, 1997).

2. See, for example, Gillian Beer's *Darwin's Plots* (Cambridge: Cambridge University Press, 2000); Tina Choi, *Anonymous Connections* (Ann Arbor: University of Michigan Press, 2015); Michael Tondre, *The Physics of Possibility* (Charlottesville: University of Virginia Press, 2018).

3. James Wood, "Human, All Too Inhuman," *New Republic*, July 24, 2000, 41–45, 41.

4. Ibid.

5. Wood, "Human, All Too Inhuman," dubs it "hysterical realism": "The conventions of realism are not abolished but, on the contrary, exhausted and overworked" (41).

6. Ibid., 43.

7. Adam Smith, *Essays on Philosophical Subjects* (London: Cadell, 1795), 76.

8. Zadie Smith, *White Teeth* (New York: Random House, 2000), 101. All further references to this edition will be given in the text.

9. See Alun Anderson, "Oncomouse Released," *Nature* 336, no. 24 (November 1988): 300; Fiona Murray, "The Oncomouse That Roared: Hybrid Exchange Strategies as a Source of Distinction at the Boundary of Overlapping Institutions," *American Journal of Sociology* 116, no. 2 (September 2010): 341–88.

10. See the excellent recent study of *White Teeth* in Josie Gill, *Biofictions* (London: Bloomsbury, 2020). Citing Lionel Trilling's account of E. M. Forster's novels, Gill asserts an analogy between the experiment and the novel as forms or practices: "Smith imagines the manoeuvring of her characters through plot as a scientific experiment" (96).

11. Daniel Defoe, *The Storm*, ed. Richard Hamblyn (New York: Penguin, 2005), 14.

12. "Both storms affected broadly similar areas of Southern England causing extensive structural damage to good quality buildings and bringing down vast numbers of trees. Indeed, the 1703 storms felled 4000 oak trees in the New Forest alone, brought down around 2000 huge chimney stacks in the City of London and even blew the roof off the Palace of Westminster" ("The Great Storm of 1987," https://www.metoffice.gov.uk/weather/learn-about/weather/case-studies/great-storm).

13. Isaac Newton, *Principia*, trans. Andrew Motte, 3 vols. (London: Symond, 1803), 2:160.

INDEX

ACKNOWLEDGMENTS

In the course of working on this book, I've benefited from the generosity of so many people. It is truly a pleasure to look back now and recall all these encounters, exchanges, and conversations that continued even during the course of a global pandemic. I'm deeply grateful to my family, first of all. My institutional home at Wellesley has generously supported my research, and I continue to benefit from wonderful students and colleagues. Thank you to Kate Erickson and Rachna Fruchbom for taking the time to talk with me about plot when this project was at its earliest stage. My amazing writing group, John Plotz, Leah Price, Amanda Claybaugh, Deidre Lynch, Theo Davis, and, for one very special year, Margaret Cohen, has been an unfailing source of inspiration, meticulous guidance, warm encouragement, laughter, walks, food, and everything else that friends can provide. Thanks also to Martin Puchner for kind advice and for the title of this book. I was lucky enough to share parts of this project in numerous talks, symposia, and seminars, including at MLA, ACLA, NASSR, Narrative, and the Society for Novel Studies; at Harvard, Berkeley, Brown, Hopkins, and Yonsei. I'm grateful for the thoughtful contributions, feedback, and questions from seminar participants and lecture audiences, and for the hospitality and support of the Berkeley English Department, the Cogut Institute for the Humanities, and the Brown University English Department, the Mahindra Humanities Center at Harvard, the Yonsei University English Department, and the Johns Hopkins English Department, with special thanks to Doug Mao, Mary Favret, Sharon Achinstein, and Andrew Miller. I am incredibly fortunate to be part of a vibrant novel and novel-theory community that includes Ayelet Ben-Yishai, Tim Bewes, Ti Bodenheimer, Marshall Brown, Stuart Burrows, Alex Creighton, Pardis Dabashi, Jed Esty, Penny Fielding, Elaine Freedgood, Matthew Garrett, Danny Hack, Ivan Kreilkamp, Sanjay Krishnan, David Kurnick, Wendy Anne Lee, Heather Love, Tara Menon, Paul Nadal, Ben Parker, Ruth Perry, Adela Pinch, Kent Puckett, Ato Quayson, Kelly Rich, Jonathan Sachs, Vanessa Smith, Ramie Targoff, Katie Trumpener, Alex Woloch, Wendy

Xin, Ruth Yeazell—thank you all so much for talking and listening, for reading drafts and proposals, for sharing and suggesting things, for encouraging, inspiring, and inviting me, for hanging out, and, latterly, reading novels like *Clarissa* together over zoom. Particular gratitude to Ian Duncan for inspiration, camaraderie, and support in many different forms over the years, and to my old friend, John Farrell. There are so many other Romanticists whose encouragement and example I also wish to gratefully acknowledge: Ian Balfour, Miranda Burgess, Jim Chandler, Alex Dick, Frances Ferguson, Ina Ferris, Billy Galperin, Kevis Goodman, Evan Gottlieb, Steve Goldsmith, Sonia Hofkosh, Jacques Khalip, Celeste Langan, Carmen Mathes, Maureen McLane, Laura Quinney, Ann Rowland, Chuck Rzepka, Kyung-Sook Shin, Orrin Wang, Matthew Wickman. Thank you to the narratology community, to Brian Richardson for generously sharing his expertise on plot, and particularly to my friend, Sue Lanser, who has done more to bring this book to completion than she will ever admit. I also wish to thank Henry Turner, Mary Crane, Jerry Singerman, Jenny Tan, Lily Palladino, Jim Chandler a second time for his reader's report, the anonymous reader for the press, as well as the readers and editors who offered helpful feedback on material in this book that originally appeared within the articles listed below.

Yoon Sun Lee, "Austen's Swarms and Plots," *European Romantic Review* 30, no. 3 (2019): 307–14. Copyright 2019, Taylor and Francis. Reprinted by permission of the publisher.

Yoon Sun Lee, "Vection, Vertigo, and the Historical Novel," *Novel: A Forum on Fiction* 52, no. 2 (2019): 179–99. Copyright 2019, Duke University Press. All rights reserved. Republished by permission of the publisher.

Yoon Sun Lee, "Bad Plots and Objectivity in Maria Edgeworth," *Representations* 139 (2017): 35–49. Copyright 2017, The Regents of the University of California.

Yoon Sun Lee, "Radcliffe's Materiality," originally published electronically in *Romantic Materialities* on the *Romantic Circles Praxis* website, http://www.rc.umd.edu/praxis/materialities/index.html.

Yoon Sun Lee, "Austen's Scale-Making," *Studies in Romanticism* 52, no. 2 (Summer 2013): 171–95. Copyright 2013, Trustees of Boston University. Published with permission of Johns Hopkins University Press.

CPSIA information can be obtained
at www.ICGtesting.com
Printed in the USA
JSHW031059060822
29000JS00001B/1

9 781512 823400